LATITUDES

LATITUDES

Encounters with a Changing Planet

Jean McNeil

BARB
ICAN
PRESS

Published by Barbican Press, Los Angeles and London

Registered UK office: 1 Ashenden Road, London E5 0DP
US office: 1032 19th Street, Unit 2, Santa Monica CA90403
www.barbicanpress.com
@barbicanpress1
A CIP catalogue for this book is available from the British Library
Distributed in North America by Publishers Group West
ISBN: 9781909954113
eISBN: 9781909954137

Biography

Jean McNeil is originally from Nova Scotia, Canada. She has published fifteen books, spanning fiction, memoir, poetry, essays and travel. Her work has been shortlisted for the Governor General's Award for Fiction, the Journey Prize for Short Fiction, the Elizabeth Jolley Prize, the Canadian Broadcasting Corporation literary awards (twice) and the Pushcart Prize. She has twice won the Prism International Prize, once for short fiction and again for creative non-fiction. Her account of being writer-in-residence with the British Antarctic Survey in Antarctica, *Ice Diaries*, won both the Adventure Travel and Grand Prize at the Banff Mountain Film and Book Festival in 2016. Her most recent novel, *Day for Night*, was awarded the gold medal in the literary fiction category of the Independent Publishers Awards in the US in 2022.

She has been writer in residence with the British Antarctic Survey in Antarctica, with the Natural Environment Research Council in Greenland, and has undertaken official residencies in the Falkland Islands and in the Svalbard archipelago in the Arctic. For the past 15 years she has lived for part of the year in South Africa and Kenya, where she is a trained safari guide. McNeil is Professor and Director of the Creative Writing programme at the University of East Anglia and lives in London.

Previous Books by Jean McNeil

From the Library of Graham Greene

Hunting Down Home

The Rough Guide to Costa Rica

Nights in a Foreign Country

Private View

The Interpreter of Silences

The Ice Lovers

Night Orders: Poems from Antarctica and the Arctic

Ice Diaries: an Antarctic Memoir

The Dhow House

Fire on the Mountain

Day for Night: Landscapes of Walter Benjamin

Day for Night (novel)

Contents

Prologue

Write the dream the character has the night before the story starts. My students consider the exercise too wayward and Jungian, they think it is a waste of time to write passages that will not end up in the published story. I tell them to trust me, there is a point to writing even when what you produce is not destined for an outcome. It's a generational difference; I was always interested in the process, rather than results.

I seem to only have anxiety dreams: about missing flights or forgetting my passport or not being able to locate a PowerPoint presentation as two hundred people shift restlessly in their seats, about embarrassments real or supposed or full-on nightmares inhabited by boa constrictors and daemons.

In this one, I am floating in a substance. It should be water, or the ocean, but it is rough and scratchy, also cloying, like polyurethane sheets. The ocean (if that is what it is) has thickened to taffeta. As in waking life, in the dream there are three voices in my head. One is me, or at least what I think of as me – the voice of my consciousness, the voice I trust. Another is the overlord: a stern, seigneurial voice that speaks to me in the second person and sometimes tells me what to do: *Write this damned book before someone else has the same idea.* Despite the form of address, I think this voice is also me, or part of me.

The third voice is external. It speaks to me rarely. In fact, I have only heard it a few times in my life: in childhood, growing up on Cape

Breton Island in Canada, in the Antarctic, the Arctic, and once, memorably, it spoke to me in Cape Town, South Africa, commanding me to do something which undoubtedly saved my life. I don't know what or who this voice is, but I think it has something to do with the land. Either the land is generating it, or it lives in the land. I also think many people have heard this voice, maybe all the people who ever lived, until very recently when we have clustered ourselves in cities where the land cannot reach us. But really I don't know.

I decide to find out via the typical writer's solution: write a book. By doing this I will uncover the identity of the speaker, this coy but commanding entity which reveals itself so rarely. The fact that I live with a voice whose origin and shape I do not know should disturb me, but I don't think I am delusional or schizophrenic, only open to a frequency that is trying to communicate with us, all the time, but to which we are largely deaf.

There is another reason I start to write this book, one which is vocational rather than metaphysical. Writing about place as an organising principle used to be called 'travel writing', even if there was sometimes no travel involved. As a label, this is now deemed to have a colonial tang, or at the very least it denotes privilege, so a new taxonomy must be found. In recent years a term has emerged, 'place-based writing'. 'Place' is a dispassionate word, but at least it takes in all kinds of physical realities – cities, landscapes, countries, even the open ocean. Still, it would be good if writing could just be writing, if new categories of consumption did not have to be constantly generated.

This book has two timeframes: 2018 until 2022, and secondly ranging over the last three decades and the wild places I have written about. The book will be an account of all years I have been living in a thermometer, watching the temperature rocket up and up with no terminus in sight. As one frame of the book runs forward the other will run backward, but in a meandering way, excavating through the thinnest crust of geological time, the dusting that is a human life.

Lately I have been having a new kind of dream, one stocked with animals, some of which I have seen and touched with my own hands and some unknown to me: moose, vaquita porpoises, wolves, giraffes. Sometimes I see only their eyes, which peer at me, fierce, admonitory. They have a mission for me. They want me to help them exist, I realise, or no, they want me to help them not die. I can't let them down. But I do. I fly six or seven times a year. I organise my life around travel and research, around impossible experiences with difficult people I know I am going to hate but I drive myself through them anyway, because I don't want to know anything more about people, I only want to learn about nature, landscape, geology, wilderness, but all my attempts to understand this realm take me back to the monster in the maze: us.

Last night the deer-giraffes returned with their liquid, chastising gaze. They know I am about to fail them. I return to the realm of the taffeta sea, the dire inky ocean. Belatedly I realise this is not a dream, this is not even a symbolic realm. It is who we are. Who I am.

PART I

WE WALKED OUT OF
THIS LAND

July 27th, 2018. Cley, Norfolk, UK. 33°C

'So, what will it be? Slimline Schweppes or Fever Tree Lite?'

Malcolm always has three types of tonic and six flavours of gin (for the record, rhubarb and ginger, raspberry, sloe, elderflower, lemon, bilberry). His favourite tonic is Fever Tree, he tells me.

'The picture on the tin is all wrong.' I point to the cypress-like silhouette. 'A fever tree is tall and anorexic-looking. Acacia xanthophloea.'

'Acacia what? Ice Floe?' His eyes already have a G&T gleam.

'Xanthophloea means white blood, or properly white tissue.'

'Trust you to know that. You know everything.' He smiles to show he means no harm.

I have been invited to Malcolm's country house in Cley, Norfolk. The London summer has turned uncanny, relentless. July 27th, the day of my journey to Cley, marks the fortieth straight day of a heatwave. The grass is tobacco. Jets scrape across glassy, un-English skies.

I take the train from London to Norwich, a journey I try never to do in the summer because it reminds me of work, with all its tensions and exhaustions. As the train sleeks through the flatlands of Suffolk I realise I have internalised this journey, which I must have done thousands of times, so that at any given moment I know where we are without looking out the window. I might know every stalk of rapeseed, every tilt of the horizontal conspiracy between land and sky.

I collect a hire car. A thunderstorm erupts – the first rain in five weeks. Soon I am driving through the villages that hug the Norfolk coast, through salt marshes rippled with sea purslane, the air laced with negative ions and their metal smell. The sea emerges. There is a lonely pact between sea and sky here, a note of displacement, as if the coast belonged to another entity, not England at all, and was trying to get back home.

The thunderstorm floats away to the east. The sun resumes its high arc but minus its heat. Now it is a tangerine smear in a cooling sky.

I feel apprehensive. I am unsuited to country house weekends; too much proximity with people I would never socialise with, too much enforced bonhomie. I am going because… my inner storyteller falters. I don't know why. Curiosity, and perhaps an incipient boredom-that-is-not-boredom but something else, masquerading as it. A flatness. That's it: I am middle-aged now and flatlining through life. All things that were going to happen have done, and at some unperceived moment I've drifted into the modals of lost opportunities, as grammarians call them. Should have, would have, could have.

Malcolm is an exporter of high-grade farming equipment with a taste for collecting art, which is how I met him, at an art fair in Paris. We started talking in a café queue and discovered we know a few people in common. He is a gregarious bon vivant, the sort of character Thackeray would have skewered, but he is not pompous or entitled. Perhaps because he is a remnant of empire, brought up in southern Africa on a tobacco farm, raised barefoot and wild until he was trussed off to the inevitable colonial child-hood cancellation of boarding school. He is cheerfully apolitical, by his own description. I tell him it isn't possible to be apolitical or perhaps only if you are very rich – which he is, 'although not really,' he is quick to say, 'I don't employ governesses and own four masted schooners crewed by young Trustafarians sailing the coast of Montenegro.' It's reassuring to know there is still a hierarchy, even at his level, I say.

Malcolm's house has a commanding view over the saltmarshes on the edge of the village. As we drink our G&Ts we watch the sun stall in the high summer sky on his terrace. The company for the weekend reveal themselves: a young Pakistani-British trader who shares

4

Malcom's penchant for Kazakh art; an interior designer to oligarchs, a tall, thin man improbably from Regina, Canada; the requisite tech start-up millionaire and a right-wing environmental consultant (they exist).

The only thing we have in common is Malcolm. He has told the others the subject of some of my books.

'Mind you,' Malcom says, before he has even cracked open the tonic, 'I don't believe in climate change.'

'It's not a religion,' I say. 'And by the way,' I add for good measure, 'facts don't care about your feelings.'

He stands back. One of the many traits of Malcolm's I appreciate: he can take riposte. He calls it 'fighting back'. I say, 'This is not fighting back, this is a delivery of the truth.'

Malcolm does not really think climate change either is not real or is real but over-blown, is my theory. He understands quite well what is happening. But he does not like the fact that there is a force in the world greater than him and indifferent to his desires and the forces that have made his fortune.

'Dinner is seeerrveed,' Malcolm cries, in his bugle-horn voice. We troop to his upstairs living room. Outside, far out to sea but visible on the horizon, is a phalanx of windfarms. They wave their arms at us, cartwheeling over and over.

For a moment I savour it, the suspension of midsummer. I love the British Isles for three months of the year. The rest of the time they are a marathon of darkness, or impending darkness.

After dinner I go back out onto his deck, as if summoned by a spell. The air has acquired a hem of chill. The sky is platinum. Within it is a taut moon, as if suspended on a string. Cries of ducks and waders float upward from the marsh. An olive sheen wraps itself around us. The red eyes of the windfarms blink in the distance.

This landscape should offer awe and succour. I try hard to feel these things, but an unidentifiable feeling, part remorse, part dread, has settled inside me. I know the reason, or one of them: the summer has been too dry for too long. Reality – or what I take to be reality – has recently acquired an uncanny gleam: bumble bees in November, short-sleeves in March, birds that decline to migrate, a soupy insistence to the normally businesslike rain of the British Isles.

Malcolm emerges from the house. 'What are you doing out here all on your own?'

I don't tell him what I am really thinking, because I can't articulate it. I reach for something more acceptable. 'I'm homesick, but I don't know for where. Kenya, maybe.'

'You know, I often think how it looks like Kenya,' he says. Malcolm has been once, on safari, he tells me, and its flax savannahs stayed with him.

We stare at the yellow seam the sedge sews into the North Sea. In east or southern Africa I would be able to identify the vegetation with a glance: Acacia-Commiphora savannah, a stand of mint-coloured fever trees, the exclamation points of marula and mopane. But strangely enough here in England, my home, I have never learnt its trees, grasses and flowers. It is as if I am preparing not to commit to this landscape, after thirty years of living in and on it.

I hear a low hum. The sound of a place, but also somewhere else. I can't describe it except to say it sounds like an echo, also a reminder, also a memory. And that it has something to do with the pulse I some-times hear, a low pulpy thud, and which is not my own.

Suddenly I know what it is. I say, almost under my breath, 'papyrus swamp'. Malcolm gives me a quizzical look, but I am already gone.

Laikipia, 0.2254°N, 37.4409°E

Ephemeral

August 12ᵗʰ, 2012. Lewa Downs, Kenya. 11°C

A vehicle wends its way below in a lazy serpentine, wafting back and forth in a manner that anywhere else would cause it to be pulled over for drunk driving. Gleaming, polished almost to smooth obsidian, it has a single driver swathed in a red cape. Behind him, on elevated seats, several hunched figures perch.

Parked on a hillside above, we watch the vehicle sashay back and forth across the sage savannah. We all know what its driver-guide is doing: scouring the bush for any sign of life, even an impala will do. Meanwhile the early-morning amphitheatre of Lewa Downs levers itself into the day like a ruined Roman city rising from the ground. This landscape has a reconstitutive element; it seems to have the power to resurrect the day before, and the day before that, and present them as new.

I start the engine to our battered Landcruiser. Eventually we sidle up to each other, the Conservancy vehicle's engine purring demurely. At the wheel of the Reserve vehicle is a Samburu guide wrapped in a *shuka* – a Maasai wool blanket with a tartan-like pattern. Out of the corner of my eye I spot the three figures in the back, rugged up against the bite of the winter morning.

'Habari ya asubuhi,' the driver greets us cheerily – meaning, literally, what is the news of the morning? Three pairs of eyes peer at us. From their perspective the news was that some weirdos have just pulled up, in the form of a not very convincingly safari-guide-like woman at the wheel and behind her (me) a huddle of seven people bristling with binoculars and bird books.

Beside me is Max, our genial if deliberately eccentric instructor. Max cuts an arresting figure. He wears a poncho of golden, Zulu-patterned fabric. On his head he has fashioned a sort of turban from a scarf. This construction is held in place with 1930s aviator goggles clamped over his eyes – a sensible adaptation for sitting all day in windy vehicles on dusty roads, although it makes him look like an escaped extra from *Lawrence of Arabia*.

The Americans in the back of the Conservancy Land Rover eventually eke out a 'hi' and our vehicles draw away from each other, two small ships pulling away into an inland sea of grass.

'They look really unhappy,' Wanjuri, who is Kenyan and from Nakuru, says. She is the first person in her family to show any interest in the wilderness, she tells us. Her father sells sofas and her mother is an accountant. They are perplexed by her passion for the wilderness. Wanjuri is paying for the course from savings accumulated working in a mobile phone shop, because her parents refuse to pay for anything other than a law or medicine course. 'If they are that miserable why do they even come on safari?' she asks.

'You would wonder, wouldn't you,' Max agrees. 'These Americans or Europeans fly halfway around the world and spend a thousand pounds a day to see some of the most endangered animals on the planet and they look as if someone has just pissed in their coffee.'

'So what's your theory?' Richard asks. He has been in advertising for most of his career and this question is his stock response to any dilemma. He lobs it in our direction at least once a day, as if we are his on-tap focus group.

'My theory is they're not too keen on the reality of the bush,' Max says, raising his binoculars. 'It's cold, it's uncomfortable. Most of the time not much happens and when something does, it's usually gruesome. Now what do we have here?' Max has spotted a vast herd of elephant stationed at equal intervals across the horizon, like an aqueduct.

Max is a one-off. He keeps warthogs and baby bushbuck as pets at home and tries to hide his chain-smoking, taking furtive puffs of Marlboro Lights and then stubbing them out, half-smoked. The only way we can tell when he is displeased is his eyelids lower to half-mast as he gives us hooded, wary looks. At night he puts on Air Crash Investigation DVDs for us to watch, as entertainment. His favourite is helicopter crashes. We learn that when something goes wrong on a fixed-wing aircraft, physics gives you a chance. Helicopters, on the other hand, go down like stones.

Lewa Downs is a large Conservancy, meaning a community-run conservation area, located on Kenya's Laikipia plateau, around four hours north of Nairobi by road. I arrived there after doing the first part of my professional safari guide training in South Africa. I came to Kenya to study east African flora and fauna, and because I'd always wanted to see the landscape that had spawned modern humans. In my training courses in South Africa, I'd spent weeks in different bush camps with instructors, all of them men. The instructors were also all white, while the students were a mix of black and white Africans, along with a few 'internationals' – people who had come from Europe or North America to do the qualification as an adventure.

'Where are you from, again?' the instructors would ask me, frowning at my accent. In safari guiding it was better to be African, of whatever race; it gave you more legitimacy, even if you'd grown up in a shopping mall-studded suburb of Cape Town. I passed for a South African, having lived there off and on for eight years and having the right to live and work in the country.

Our instructors had worked in the industry for decades. They were thick-kneed and phlegmatic to the point of muteness. They reminded me of the war correspondents and soldiers I've met. No violence would animate or disappoint them. 'That's the way of nature,' they would say, the flat note of fate in their voices as we students covered our eyes or looked the other way as a lion took down a waterbuck or wild dog ripped apart a hare.

At Lewa Max floats between us, trailed by mauve ribbons of Marlboro Light, his eyes drinking in the horizon. 'Beautiful, so beautiful,' he mutters, to no-one in particular. We students sit around campfire under the olivine skies of evening. With each day we ramble around the Conservancy, set high on a fertile plateau soaked in volcanic soil, its hills undulating smokily in pulsating layers of beige, umber, flax. Mount Kenya presides over us, its serrated peak streaked with ice streams while on the palomino plains below it is twenty-eight degrees.

But here is the strange thing: now we are living in the middle of its splendour we can't appreciate it – Richard and I at least are aware of this paradox. So quickly, we comment to each other, the human mind normalises your surroundings. We start out marvelling at the striped kingfisher in the fever tree calling, or the whipping inner-tube form of a spitting cobra flashing through the grass. But then the land ceases to throb with metaphor and your thoughts turn to what is for lunch: salad, hopefully, to leaven our gruelling routine of *ugali* (maize meal) and spinach.

'Let me tell you,' Max drawls, his eyelids hung over sage green eyes at half mast, 'this landscape knows everything. It knows you are ephemeral. So what do I mean by that word?' Max asks us. Everyone's eyes swivel toward me. I have told them nothing about my true identity, but somehow they suspect I know the answer.

The word has stalled in my mind. *Ephemeral.* I am thinking about how I have always loved the geometrical lean in its sound, how this landscape broadcasts its insistence on existing in a way I have never felt before. Like a slab, or a hand pushing me toward a precipice. At the same time it feels as if we are on an ocean liner, as if the plains of Laikipia are mobile and floating through time, but our point of embarkation and destination have been forgotten.

I look up to find the group still looking at me expectantly.

'Transient,' I say. 'Fleeting, passing, ghostly. Something that lives for a very short time.'

Running

We shuffle listlessly, standing under a thorn tree on a cold Lewa morning.

'Once the human being has left the Land Rover – or Landcruiser, let's not play vehicle favourites – what advantages does it have in the African wilderness?'

We trade glances: *another of Max's trick questions!*

'Battery-operated torches?' proffers Richard.

'Botanical gin?' I say.

Max gives us that seemingly neutral stare with his khaki eyes and turns away.

'I guess we didn't pass that test,' Richard hisses.

The punch line to Max's question is: none. Lewa Downs has a healthy population of lion and leopard. In a deep chamber of both species' limbic brains, a switch sits on a dark wall. It is either in the on or off position. When on, the animal becomes a machine for pursuit. No human has a chance of escaping it. The first maxim of bushwalking is, whatever you do, don't run. But it is a hard instinct to smother.

In London I run eight kilometres a day. At Lewa, galled on by my endorphin addiction, I run circuits around the perimeter of the camp while my fellow students sit in camp chairs drinking tea and casting me narrow looks: *we're walking for seven hours a day, is she doing this to make us look bad?*

After a while I devise a route that allows me to theoretically jump into a tent, an outhouse or vehicle within five strides. But what would I really do if a leopard appeared, bursting onto the road with its coiled gait, its tail cobra-ing back and forth? Would I be able to move fast enough to outwit this arch ambush predator?

I tuck my hair into my baseball cap so that I do not do quite such a convincing impression of a female impala with her bushy tail. Another problem with running in this environment is you can't scan the near horizon for tawny lumps of muscle moving at twice the speed of top racehorses, because you have to look down. The reason: snakes.

Specifically, the puff adder. Common throughout east and southern Africa, the puff adder strikes at the same speed as a Boeing 777 departing the runway (200km/h). Even slow-motion cameras can't record the velocity of their strike. Alone among snakes, adders do not react to human footfall. They stay still until you are too close to avoid them. When I ask Max what drives this complacency he gives me a long, steady look. 'From a puff adder's point of view, they're going to kill us before we tread on them, so what's the point of moving?'

One day as I orbited camp I spotted three young lions around two hundred metres away. They yawned and entwined themselves around a log. They stared in my direction with that amber implacable gaze of theirs, no doubt perplexed by the weird circuit-training antelope huffing by.

But their state could change quickly, were that chase-hunger switch to be flicked. Lion attack at around 88km/h, top speed. Leopard are even faster, at around 92km/h. It would be like being hit by a speeding police car – what in fact gave me my most convincing brush with death, twenty years ago now, on Kingsland Road in Dalston, north London. Would I even see them coming? On the edge of my eye I might see a haze of taupe, the bread-loaf outlines of the lion's quadriceps. A plate-sized paw would remove my face in half of a second.

Along with a complex language and abstract thought, the only advantage humans have over a leopard or a lion is the concept of inside. Tip for fleeing a lion – find a tent, car, or other closed structure, preferably fast. For the lion it will be as if you have vanished into thin air. As I huffed I measured the strides I would need to make it to the long-drop loo, one of my tents or our Landcruiser (nobody locks cars in the bush so you can just open the door and throw yourself inside) or the shipping container we used as a storeroom. Three, six, nine. A lion would cover the distance in a flash.

I kept running. Occasionally the young lions would throw me a look of bored appreciation, the sort of glance you might give an attractive stranger in an airport.

Why wasn't I afraid, or *more* afraid? This question dangled in my mind every day since I had started to learn to guide. *This landscape is more powerful than you. Why are you daydreaming about the enduring influence of Weimar Classicism on contemporary fiction when you could be about to be gored by a buffalo?* That was the limbic inner drill-sergeant, trying to keep me alive in a place where there were at least five species of wild animal that could swipe me off the planet, but which I hoped would be too busy occupied with eating or mating to bother.

There is another reason I keep a tight rein on fear. Animals smell it. Over on the log, the young lions licked their paws, drawing their sandpaper tongues lovingly over the Australia-shaped pads, meticulous and slow, as if they never intended to use them to fell quivering Grant's gazelle, to rip tendons from muscle like cutting a string-bound package. I thought again about the brutal proprietorial *mien* of predators, how certain they are of their right to destroy.

Running is very good for thinking. Apparently, the lateral eye movement required to keep yourself straight on a course and the rhythm of the unfolding limbs can alter brainwaves. When running we waft into a meditative frequency, seemingly, but actually several higher brain functions are ignited by the simple task of putting one foot in front of the other.

When running the land also looks more eternal and serene. What I see: oceans of panicum grass, floating in the wind. The red-purple undulations of *rooigras* shimmer, terrestrial mirages, in the sunset. Altocumulus clouds preside, casting airship shadows across the land. Wedges of light, dark, light, dark float across the grasslands into pools of flaxen sedge, coaxing the doom of afternoon thunderstorms. Raindrops hiss and spit on our campfires and we take refuge under the lintel of our tents. Fever trees stalk the waterways with their green-gold poise, like willowy teenagers. When the sun finds them they glow a gold so thick it becomes almost blue.

At Lewa I have arrived at a rare internal amnesty, I realise. I am neither running toward something nor running away. I feel at home in the world, protected, even, by the serene magnetism of Mount Kenya, which we can see from our camp. Its serrated shape, the way it seems to both rip the sky open and merge into it, reminds me of something and for days I struggle to place what it is. On one of my runs around camp it comes to me. In Norse mythology, the realm of the Gods is called Asgard, a halcyon zone presided over by a mountain, Gimle, which translates as 'golden-roofed place' and also, 'the place safe from fire'; a pagan heaven.

Perhaps I feel at peace because for once I am not in a place to write about it. My bizarre quest to train as a professional safari guide has nothing to do with writing, I tell myself. On the course I keep a diary, full of metaphysical treats such as the average weight of buffalo (700-800kg), the six most common wildlife diseases (number one: Anthrax), or the breeding behaviour of Sable antelope (agitated).

Still, it is hard not to want to write about Lewa, to try to capture its brooding, amber glamour. Like most writers I am a devotee of a cult called Get it on the Page, trying to tether reality to my consciousness, rather than perform an act of homage. How can I get beyond the cool, brooding margins of this place and its tobacco hills? I realise that if I want to write about this place, the far field of its *there*-ness, I will have to break free of language.

The Edge of Reality

We arrived back at camp after a cooling afternoon spent outrunning the front edge of a rainstorm. We sat in a semi-circle, each of us lost in thought, cradling Tusker beers on our abdomens, mesmerised as usual by what Max called the 'bushveld television' – our campfire.

I looked up to find Richard peering at me. 'Why are you doing this?' he asked.

I'd first overheard Richard at the tented camp where we were all billeted in Nairobi before the course began. I was fascinated by his accent. It was impossible to place, a fusion of American and English: or rather the vowels and cadence was American, but there was a shallowness to his 'a's and 'e's, and the occasional glottal stop.

As it turned out, he had once been English in an alternate dimension known as the 1980s, before emigrating to the States. He had retained the Englishman's reserve about money so it was hard to tell, but it seemed he might have made a small fortune. The signs were there in his references: house in upstate New York, winter house in Martinique, perma-tan, Swarovksi binoculars. He described himself as a socialist. 'I'm a lifelong Labour voter,' he'd announced to me, as if presenting his credentials. 'Me too,' I'd said, 'pity they aren't in power more often than twice every hundred years.'

Richard was waiting for my response. He had caught me off guard. Maybe I'm reckless, I could say. Maybe I'm depressed. Certainly I was a lot poorer than if I hadn't flown myself to Kenya for two months and signed up to an East African safari guiding course.

'I usually don't know why I do things,' I said. 'How about you?'

Richard had spent a lot of time in Africa, he said, undertaking expensive trips in helicopters or on horseback. He explained that he wanted the qualification so that he had the legitimacy, in the eye of his clients, to lead safaris. 'I want to be able to say I'm a trained guide.'

'But you're very experienced. You could say that now.'

'I want it to be objectively true,' he said.

Max came within earshot. He was trying to smoke out of range of us students and failing. The wind filled our nostrils with Marlboro Light.

'What are you two talking about?'

'The difference between objective and subjective truth,' I said.

'Nice. Do you know how to tell the difference between a caracal and an African wild cat track yet?'

Richard's question opened a fissure. We are called to account for our motivations all the time now, but anything I say would be dishonest, in that way that self-disclosure is never honest, until it is hewn into literature.

I am 44 when I begin training to be a professional safari guide. I am leery of giving my age away to my co-students or to Max. They seem to assume I am younger. And if they knew the truth? It might coax the voracious pity that men lavish on women: woman with no children and husband. Woman unmoored from her life, looking for direction, panicking, even, as she watches her future be overtaken by her past. Middle-life is a misnomer, this woman (note I have instantly cast myself in the third person, a character in this evolving story about a middle-aged woman who nurtures delusions of becoming a professional safari guide) thinks, as the wind rustles through the long grass, soothing and chastising her at once: she is more than half-way through the ride, especially if you take into consideration that all her relatives have dropped dead quite young, with the meanest living the longest – yet another example of the ironic revenge of the Gods.

The quest for knowledge is the official reason this woman is putting herself through the ordeal of qualifying as a professional safari guide. But underneath this runs a dank current of something she calls *lastness*. She wants not only to be able to walk anywhere in the continental African wilderness south of the Sahel and survive, but to be able to interpret it, to write its story as she might plot one of her novels. By learning to move through it she will become a walking diorama, an archive of impressions. Qualifying to be a safari guide is a kind of commemoration, she thinks: her project is to know this realm intimately, before we extinguish it.

There is also a personal quest, as there must always be, in any story. This woman wants to cleave herself in two in order to solder together another kind of person: the economic actor self who holds down a university job in an increasingly draconian higher education sector and who can quote passages of Alain Robbe-Grillet verbatim, versus the self who is entranced by the natural world and is trying to find a way to indulge her enchantment without bankrupting herself.

In the bush she can live in rapt attention to the now, and thus in a state of suspended animation very like youth. For once this woman is not beholden to screens and devices, to the demands of being just another unit of labour, neoliberal cannon fodder, censured and trimmed at every stage by credit card companies and insurance companies and flight consolidators and electricity providers that feed upon her body. She can resume her ancient position, abandoned long ago, 50,000 or 500,000 million years ago – what does it matter? – of being human. A human being.

*

Four weeks into our course, our written exams approach. These are widely considered difficult to pass. We need to get seventy-five percent or more to obtain our qualification.

'There are six hundred species of grass in Kenya alone so good luck with that!' Max's laugh somehow manages to be encouraging rather than caustic. He teaches us their Game-of-Thrones names: *Aristida, Chloris, Cenchrus, Chrysopogon, Exotheca, Hyparrhenia, Heteropogon,*

Loudetia, Pennisetum, Panicum, Setaria and *Themeda triandra*. We must also learn dozens of ungulate species I've never seen in southern Africa and whose names sound as if they've been made up by sleepy children: puku, bongo, hirola.

Slowly, though, the land emerges from the shadows, coaxed by knowledge. Mark teaches us the trees: the flamboyant (which originated in Madagascar), with its shocking carmine crown of flowers, a blur of acacia – *tortillis, mellifera, elatior* – the Nandi flame, so named for its flowers of a deep, throbbing orange. But these are increasingly being out-competed by invasives: famine weed, clusters of evil prickly pear, Australian blackwood, Mauritius thorn.

And then there are the birds.

'A bird is a bird,' says Richard.

'You don't know your birds, you don't get your certificate,' Max counters.

'Why can't birds just all look the same?' Richard says, to no-one in particular.

Before training to be a guide, I'd thought birdwatching was a slightly more enlightened version of trainspotting. But after nearly a year's immersion in the bush I discover I am in fact one of those middle-aged people in North Face walking gear with overpriced binoculars plastered to their eyes and clutching thick tomes with titles like *Field Guide to Fidgeting Flitting Things You Will Not Know the Names of*. 'Great job!' Max writes, in a rare compliment, when I get twenty-five out of twenty-five on the bird call test. Now I know what keeps the hardcore birdwatchers addicted. It isn't after all the list-checking but the authority, the stability and certainty in being able to say *night heron, Hartlaub's turaco, golden-breasted starling, double-banded sandgrouse*. Our brains love order and control, and behold, the 2,477 species of African birds are there to oblige us.

'To have knowledge of things one must first give them a name,' writes Jamaica Kincaid in her essay 'In History', which traces the relationship between colonialism and natural history. Naming things is reverence, she writes, but also power, particularly the power to control the so-called primitive and the wild. Naming is a refutation of chaos. No accident then that it was European men who were allowed to name the parts of the natural world – the Humboldts and the Temmincks, the Von der Deckens thought they were in virgin territory. Things in the world did not need their names yet here they are: Von der Decken's hornbill, Temmincks Courser, Humboldt's penguin, consecrated forever, or at least as long as we, and they, last. Then, the birds who bear the names of these expeditioners will revert to being their unnameable, true selves.

High on the dopamine hit of being able to name one of over 260 bird species, I could have stayed at Lewa forever, firing off *African black-headed oriole, four-coloured bushshrike, Abyssinian nightjar, malachite kingfisher.* Names of crushed emeralds and medieval armour, like arrows in the night or leaves facing the dying light of the afternoon, bugle trills and pineapples made of soldiered metal. The black and Klaas' cuckoo had such plangent calls that when I heard them it was an effort not to burst into tears on the spot.

Naming is also a form of murder, as Kincaid concludes in her essay. You have to kill the thing named, chop it up into pieces, then resurrect it in a different form. Writing is exactly like this, too. Mimesis and exposition and other strategies of representation require this grisly dismemberment. The whole aesthetics of fiction rely upon creating a realer-than-real reality. The problem is, real reality – like Lewa – effortlessly resists representation, because it is so outlandishly real it defies human consciousness, which after all only seeks to claim it for itself. It is not a stretch to think that the land, as opposed to humans, has a completely distinct reality, one that defies time, circumstance, and certainly anything as tawdry as cause and effect. Our perception can only run along its edge, as if balancing on a narrow ledge above a precipice. We can only pick up is a vibration, a low restless hum not far from a tune. When I hear it, all I can say is that it sounds like a song once familiar that loops us back into a dimension we once knew but have now fiercely forgotten.

The night before our exams I can't sleep. The large prides of lion who roam the Conservancy seem to delight in killing things within our ear-shot. The squeal of a kudu being garrotted is not, as it turns out, a sed-ative. I try earplugs, then, in increasing desperation, my iPod Shuffle, pumping Cinematic Orchestra into my head to paint over the kudu's ordeal. What I really want to do is get up, fetch a rifle, and shoot the giant tawny leeches which have attached themselves to the kudu's neck.

Eventually the night-time kudu garotte-fest and the stress of imbibing stacks of information brings me close to delirium. My eyes jolt open at 2am trying to remember the Latin name of a knobthorn tree (*Senegalia nigrescens*) or the name of the middle star in Orion's belt (Alnilam). In those long hours when I lie awake, every sense on alert, it occurs to me that by becoming a professional safari guide I think I will fix something about myself. I want to be dogged and capable. I want not only to be able to interpret the wild, but to *be* the wild, to take something of its power and indifference into me and displace the person I have become, drunk on ambivalence, obsessed with the emotional pointillism of being a writer.

That night I do actually get to sleep. But around three in the morning I am woken by a quiet commotion. I peer out of the flap into a moonless night. A galloping sound envelops the camp, as if generated by the land. As my eyes adjust to the darkness I see what looks like hundreds of green fireflies flying through the night. A herd of impala are on the move, right in front of my tent, perhaps spooked by a lion. Their russet bodies chevron over the land as they perform faultless arabesques. They vault the fence that surrounds us effortlessly, one after the other, like waves on their way to some other ocean, a floating detonation in the night.

The Quenching

'I've got something special lined up for you lot today.'

We students trade sideways glances. Max has been brewing ordeals for us lately: teaching us offroad driving tactics on scary inclines of raw mud, surprising us with 'flash tests' (his term) and survival ordeals such as camping out overnight in the bush with the Conservancy's rhino protection team, who are better gunned than the Israeli army. We have just spent a night sleeping in tree branches and learning to make a fire with no matches.

'But we need to wait until the afternoon when the wind is right.' This test is meant to be pleasant, Max tells us, a relaxing interlude before our exams.

We set off on foot. Ruby, the *askari* (guard) the Conservancy has assigned to Max so that we hopefully don't get killed, accompanies us with his AR-15 resting by his thigh. ('That sorry excuse for a gun won't help us much if the shit really hits the fan, you need a bolt-action rifle with at least a .375,' Max confides to us one day. 'It's just for show.' 'Thanks for that Max,' we say, 'that's really reassuring.')

The afternoon is oddly silent. Where is the striped kingfisher, who calls in an unceasing cascade? Where are the hee-haw calls of the Grevy's zebra stallions who regard us like schoolmasters as we march past them? The air is clear and we have a good view of Mount Kenya, streaked with winter snow. The only sound is of the grass parting as we move through it, a hissing noise, as if it is scolding us.

Max directs us to a fallen tree branch. A couple of hundred metres away three of the Conservancy's white rhino graze in a glade of adolescent fever trees.

'Ok students, let's sit quietly here.' Max points to a felled fever tree. We dutifully line up on the log. In no time the rhino approach, walking shoulder to shoulder in the grass, grazing, sweeping back and forth like a cadre of Mennonites scything the fields. They keep coming until they are only ten metres from where we sit.

There are two species of rhino, black and white. Their names have nothing to do with colour (all rhinoceros are grey) but with the shape of their faces: white is widely interpreted to be a misreading by the English of the Afrikaans *weid*, meaning wide, referring to the white rhino's dinner-plate shaped mouth. The black rhino is named on the same antithesis principle as the Antarctic – for what it is not. Black rhino are smaller than the white and obviously start reading Deuteronomy and Corinthians straight out of the womb because they certainly know how to hold a grudge. We hear stories of them tracking people who have annoyed them over days, then ambushing and flattening them. The rhino in front of us are the more placid and reasonable, if substantially bigger, white version.

Up close, the rhinoceros looks as if it has been assembled from several different creatures. Its head is an oblong block of concrete studded with a 2.5kg-heavy horn. Two seed-like eyes stare out from heavy-lidded sockets, surprisingly expressive. Rolls of tough flesh fold down over the tops of its legs. Its ribs stand out like scaffolding. I have a flash of an image: a Viking longboat wreck.

We all hold our breath. Apart from Elvis, the reserve's tame rhino, this is the closest I have ever been to one while on foot. What will we do if they turn and hammer toward us? There is no way a human will outrun a rhino or avoid its battle tank bulk. The only chance you have would be to climb a tree – fast. If you did have a bolt-action rifle you'd have to land a brain shot to kill a rhino, and then the Conservancy would probably sue you.

Throughout Africa, the black rhino is critically endangered, at risk of becoming functionally extinct, while the white rhino is classed as endangered. Mine may be the last generation to see them in the wild in Africa. Next door to Lewa, at Ol Pejeta, also a Conservancy, the last three individuals of the sub-species northern white rhino languish in an anteroom to obsolescence.

The reason is that humans are hunting these modern-day Triceratops to extinction. The word comes from *exstinguere* – Latin for 'to quench'. The meaning has changed over time. An extinction was originally a cool draught on a hot day, a fire put out. Only since 1784 has it come to be used in English to mean a never-ing, an endgame. Even before voracious *Homo sapiens* showed up, extinction was the default experience for all organisms that have ever lived on earth. Over 99.9% of species that have ever lived are now extinct, as Max reminds us most days in our lecture tent. No species is guaranteed to live forever, no matter how hardy and ruthless, although some like the crocodile have played the long game well.

The air around us thickens. Now the striped kingfisher calls ceaselessly, as if anxious to deliver us a message. Black-bellied bustards coo in the grass. The wind threads through the whistling thorn, always a sombre sound, like oboes whose reeds have split. The blue afternoon slides into gold. This daily declension of afternoon into evening at Lewa is beyond a mere slipping of the light, or any natural phenomenon. It makes us wistful, too aware of the finitude of time. Just as being surrounded by wild creatures who are probably doomed, and probably very soon, creates an internal constant requiem that is impossible to drown out.

Max puts a finger to his lips. When the rhino are only ten metres away, one exhales – a sudden, fast jet of surprise. She (we decide it is a she) raises her head and looks straight at us. We have been busted. We hold our breath, watching her eye evaluate us. What does she see? A rank of outsized birds lined up on a branch. Her eye blinks. She turns away. The other two rhino obey her signal. They change course so delicately, for an animal that looks like a Panzer.

We clamber down from our branch and make our way on foot back to camp. The rhinos' force field stays with us, with its morbid gleam. And we, their quenchers, who were so close by, listening to the dry huff of their breath. Many times in my years in the African bush this would be the truth of the encounter: two animals that had once been united in the same landscape, now separated by 1.5 million years of evolution, listening to each other breathing.

Soon the word would gain a foothold in my mind, locking into it with pincers, like a tick. *Anthropocene.* An ungainly word. It will never catch on, I think. The Anthropocene is not properly speaking an era, but an open-ended terminus of the Holocene, the era we inhabit. The Holocene began around 11,650 years ago, after the last ice age retreated and *Homo sapiens* fanned out across the world. Its name is derived from the Greek: *holos* (whole); cene is from the word *kainos*, meaning new. The tail end of this Whole New era has only very recently been re-christened as the Anthropocene, the era of man.

In Kenya the colonial era began with European settlement in the 1890s. Until then, the plains of Laikipia on which we learn to guide swam with game: oceans of zebra, wildebeest, giraffe, Grant's gazelle, the quizzical Liechtenstein's hartebeest with their PlayStation horns lapped over these lands. No-one knows the exact numbers; there are no written records of the plenitude.

Sitting on a *koppie* one evening at Lewa after our daily bushwalk I imagined the scene, sketching in ranks of cheetah, sway-backing like a flotilla of question marks, feeding on acres of impala, so many the landscape turns umber under the blanket of their bodies. Elephant wend through stands of doum palm without fear. A crash (the term for the collective) of rhino gallop, the sky empty of helicopters bristling with weapons with tritium night sights.

Now the ordinary pattern of rainy and dry seasons in Kenya, around which agriculture, land cultivation and nomadic herding has developed over thousands of years, has become erratic, even bizarre. Areas in Kenya which are already dry are drier. Long droughts that parch the land and its animals to within an inch of their lives are now followed by epic floods which inundate entire communities.

Let's imagine this vast grassland plateau in 2100: what do we see? The peaty black cotton coil cracked, arid. Emerald grasslands tarred. A sere wind rustles through a graveyard of fever trees, dead from thirst. The giants of the plains, the elephant and buffalo, are vanished, replaced by clots of blood-coloured Ankole cattle that bristle on barren pasture, sparse grass crackling like glass in the wind. Mount Kenya still regards the plains with its gap-toothed grin, but no snow has fallen there, not even in the dead of the upcountry winter, in two generations. In the villages on its flanks elders tack photographs of its ice-streaked peaks on the walls. As the afternoon temperatures approach fifty degrees they take refuge in the darkness of their huts, staring at the image of their once-cold mountain.

The Trophic Pyramid

The next day there is a storm. Rainstorms at Lewa do not arrive but rather materialise out of sky-matter, as if generated spontaneously. From our camp, parked in our chairs between our Landcruisers, we watched it come. One second the sky is clear, the next it is clogged with battleships.

We are realising, slowly, that sky is not sky at all, meaning that cape that protects us from gamma rays or whatever else the sun has to throw at us. It is a dimension. It generates reality, it is a factory of simulations, of matrixes and articulated geometries. There were signs we failed to read until it was too late; the cobalt under-shade of the clouds, the way the Nubian woodpecker battered at the fever tree, as if it had wanted to be hit by lightning.

The rain is so dense it threatens to sink our lecture tent with its weight. We have to stand on chairs and drain the canvas every few minutes. There will be no aleatory walks in the tall panicum grass today. We will do what Max calls 'book-learning'. This is what takes us to Max's next test, which we will all spectacularly fail.

'So students, what is the immediate ancestor of *H. sapiens*?' The hawk note has returned to Max's gaze.

'*Homo erectus*,' I say, relieved to know something for a change.

'Good, how long ago did *H. erectus* live?'

'Forty million years,' I venture.

'Four hundred million,' Wanjuri says.

Richard has the last go. 'Four hundred thousand million.'

Max's eyelids droop to half-mast. The troughs on either side of his mouth harden.

'Didn't you pay attention in geography class? Biology? History? An eleven year old can answer that question. You call yourselves *educated*?' For the first time since our course began four weeks previously Max is genuinely upset with us.

By the end of August the upcountry winter is softening. I no longer wake in my tent at 5am to puffs of condensed breath, then have to execute my mad scramble to put on five layers in the ten-degree cold. But the coming of the equatorial spring also signals the end of our time at Lewa. Soon we will return to our ordinary lives, scattering to jobs now forgotten about, family kept in touch with by one desultory text a week.

Group endeavour, especially in the wilderness, is an unsought totalising environment, like joining the Army or a cult by mistake. In any group dynamics there are turning points. The internal hardening of our group, until this point friendly and supportive, has a reverse mirror in the bush around us. It is hard to locate when or why the tenor of our group changed. I know I had fallen out of favour with Max over the previous few days because he has discovered I have a saviour complex.

We were out walking near camp on a birding expedition when we came upon what we thought was a dead African dormouse, a common savannah-dwelling rodent. We stopped to inspect it. Its silky, golden fur shone in the slanting morning sun, filtered through a mesh of acacia leaves. The animal's large blackcurrant eyes were squeezed shut. As we circled around it, it twitched. A shriek came from the fever tree above us. We looked up to see a familiar form: a black-shouldered kite.

'Ah, the old playing dead trick,' Max said. 'The kite had its eye on the mouse before we came along. It might even have caught it and dropped it and is waiting for us to leave.'

I looked at the small creature in the grass below us, how its eyes were screwed tight against themselves. Its look of terror and dejection was familiar to me, but in a way that was watery and indistinct, like a faint signal picked up from a distant galaxy or another lifetime.

I bent down and picked up the dormouse. It seemed to weigh as much as a paperclip. It opened its eyes and gave me an inscrutable look. Probably it thought I was the kite, and its life was at an end.

'What are you doing?' Max barked.

'I'm giving it a chance.' I walked away, the dormouse held between my fingers, looking over my shoulder to make sure the kite was not watching. At that moment the bird flung itself from the branch and levered itself into the sky like a jump-jet Harrier. It flew away, a scar against the sky. I deposited the dormouse beneath a large fallen branch. It wiggled out of my fingers, shook itself, and bolted into the grass.

For the rest of the day Max would not speak to me. One of the cardinal rules of guiding is never to intervene in a situation between animals, on two grounds: one, you might get hurt. But more important is the principle that the food chain is a self-regulatory mechanism. Lion, hyena and wild dog might collectively be a hellhound, but predators need to eat in order to survive. Someone has to be killed, otherwise the trophic pyramid – the ecological dynamic that has evolved over billions of years in a particular ecosystem – would crumble.

When my eye first encountered the term in a textbook, I misread it as *Tropic* Pyramid. As in the rain of the tropics, rushing down a triangle, soaking it, sluicing. There was something juicy and alluring in the word's cascade. Max had corrected me: *trophic*, meaning related to feeding and nutrition. Still the rush in the sound of the word made me see a waterfall, a precipice we could all slide down, given the right circumstances.

'Human emotion has no role in this logic system, other than to destroy it,' Max said, by way of admonishment.

I agreed with him. I knew I had deprived the black-shouldered kite of his or her lunch, but the instinct to save the animal to run through the *rooigras* for another day was too strong. I knew the price;

Max might fail me on the course, just because he could. While he had many qualities I admired, he was one of those on-switch/off-switch people. Once you do something that goes against their code, the switch never flips back in your direction. I knew he was right in the law-of-the-jungle sense, yet I knew that given the chance and the right circumstances, I would do everything I can to save an oryx cow and her calf menaced by wild dog, or a tortoise being harassed by a jackal. My goal was not to wreak salvation, or not only, but something more furtive: atonement.

We are in the classroom and the rain whooshes down the canvas outside, so hard that Max has to shout to be heard.

'Human beings were made in this landscape,' Max continues. 'Right where we stand. Three hundred thousand years ago, we walked out of this land. This is our origin, our moment zero.' Max tells us that the fact that humans evolved alongside animals in Africa is one reason why the megafauna on the African continent was not swiftly wiped out as in New Zealand or Madagascar – there had been a long, drawn-out emergence, a co-existence during which a kind of truce was stabilised: so many hominids killed in caves by lions measured against so many zebras hunted by humans in canny bands.

But, I think, listening to Max's lecture on the coeval development of humans and animals, there is something here we are not acknowledging. Something we have no word for. Humans are not an apparition, nor a point of origin, but an enigmatic moment that is still evolving, blooming across infinity's horizon. We are not like any other animal because we are not entirely subject to time. We can project ourselves into the future, we can imagine false pasts, we live in a haze of conjecture and supposition. I have always thought that there is something uncanny about us. This does not lead to a deity, but I have come to believe that something has willed us into existence, for its own undeclared project.

I think of the phrase used by the Khoisan, the original hunter-gatherers of southern Africa, for the stars: they were *the dreamer's dream*. Somewhere beyond our planet, the dreamer dreams us into being. The dream is unfinished; it may yet become a nightmare. It is not a cliché to say that on the African continent, especially here, in the landscape of the moment zero of our species, we are closer to cracking the reality of *us*.

Later we sat around the fire, waiting for the evening flotilla of grey-crowned cranes to fly overhead, which they do at 6.08pm every day, so punctual you can set your watch by them. The embers of the fire burned to a low, hungry red. A sheen appeared on the horizon. Soon the moon would appear and pour platinum light that zig-zagged across the land in stark chevrons.

Max's face had a ruminative cast. The fire painted it ochre. For a second he appeared to have changed consistency, to have become another species entirely.

'It's like I've opened my mouth and swallowed this place, whole, or it has swallowed me,' he said. It was true: we would never evict it, all of it, the yellow-throated longclaw's clarinet sundown call, the stark acacias, the giraffe that floated like chaise longues across its tallow-coloured plains. Vast cloud-shadows stalled above us, lingering like delegates delayed on their way to an important conference. We stared and stared day and night, strafing it with the buried zeal of unrequited love, and unlike a human being it never let us down.

*

Our last week in camp was full of practical tests and exams. We would pass or fail, and within a week our small world would fold itself away as easily as the camping tables and chairs that were our only furniture.

I passed. We scattered to various cardinal points of the earth: Richard to New York, the Kenyans to Nairobi or Nakuru, the South

Africans to Joburg and Durban, me to London, Max home to his family of semi-tame warthogs. In the autumn I resumed my teaching job. On my journeys home from Norwich by train in a deepening autumnal darkness a strange depression settled inside me. It was not the ordinary emptiness of the creative writing teacher, who has to give everything to her students, digging up vast inner resources of malachite and topaz to hand over so that the students can fashion market-ready jewels. To banish it, on cold train journeys I listened to the calls of birds I had learnt at Lewa, the black cuckoo, the green-backed camaroptera, the striped kingfisher.

Lewa hovered in my mind all autumn that year, a zone of mysterious pattern. I thought of its phantoms: the *Simbakubwa*, the giant lion-hyena, of *Equatorius* and *Nacholapithecus* and all the other megafauna of the Miocene savannah and their descendants, the rhinoceros, soon to be quenched in the clench of extinction. Before I went to Lewa, I was aware that humans had a genetic home landscape, one which has withstood our onslaught since the very beginning of us. But I hadn't considered that the land could feel different, merely from having a long memory of us, as a species: more serene and authoritative than many landscapes I had lived in and written about. Lewa made me consider the land might have an ideology, one unrelated entirely to the human notion of *home*, that plangent word in which the yearn to belong is already audible. That the land could be generative only of itself. It does not need us at all. And in its indifference, we humans feel that old thrill, now unfamiliar, of abandonment.

PART II

THE BOWL OF WINTER

September 2nd, 2018. London, UK. 26°C

The long, hot summer finally draws to a close. I find myself looking forward to autumn, never my favourite season. The relentlessness of the heat this summer created a strange energy: anxiety twinned with resentment underscored by a trilling dread. Drawing the blinds every day I had the impression we were no longer living in the British Isles but in Poland or Wyoming. Winter is usually the season of forbearance but this summer required a different kind of endurance I've never encountered before. *The land can't withstand this,* I found myself thinking. *Please let it soon be over.* For me, a person brought up in the forever winters of Canada, to foreswear summer is to let go of life itself.

On the second I host a modest fiftieth birthday party with ten friends. My circle of association is narrowing, like a camera aperture tightened against the light: at thirty years old I had fifty people in my kitchen, for my fortieth birthday party, twenty. I never expected to be one of those people who hire entire houses or even ships (as Malcolm did for his seventieth) for decadal bacchanals. Still, all I have been able to gather around me, these years on the planet, is suddenly revealed: a flat, a job, the books I have written, a clutch of friends. A small harvest.

As for turning fifty, it ignites no existential crisis. In human terms I am now too old to be said to have died young, that's all.

But wait, something *has* changed. I start to have dreams of a new order. They are no longer a catalogue of predicable anxieties: arriving to teach a class unprepared, finding myself stranded in an airport. My dreams start to take on an epic tinge. In one I wake to find it is 10am but totally dark. Could the clock be wrong? Where am I? In some country – Norway? Finland? – where the sun does not rise. But even there in midwinter it is not dark this late. Something has happened to the sun, I realise. It's gone out, like a blown bulb. The darkness outside the window is strange and different to anything I've seen before, even in the Antarctic. It is solid and leering, like frozen petroleum.

After I returned from the Antarctic, I dreamt of the ice continent for nearly a decade. No other place (except the Falkland Islands, of which more later) has lodged itself so stubbornly in my subconscious. In these dreams, aircraft hangers were housed inside icebergs; to get to the plane we would have to move vast sliding walls of ice. In another, a sheer island of ice unleashed from the continent crashed into Buenos Aires and destroyed the city. In another, ice crept over the world, a reverse global warming. I looked out the window to find a glacier parked outside my flat in Stoke Newington. In the dream the ice is a giant jellyfish that covers the planet. But is it ice? The physical state of it is a decoy, as it turns out. The ice is alive, a sentient substance. The energy that moves between the ice and myself is abashed, electric, almost sexual.

At one in the morning I am still washing dishes: flutes with warm cava pooling sourly, plates encrusted with rubber-red beetroot humous. I'd forgotten about the ice dreams, until recently, just as I seem to forget most of my life. I rinse a sieve, watching the water run through it. It is this moment, fuelled by the residue of the tobacco fields of our nearly two-month-long heatwave that summer, that convinces me to try to write a documentary sort of book about being alive in the Anthropocene. Writing is the only way I will ever get to the bottom of how it feels to be alive now, its various uncannies, like being in a theatre of ghosts: to be forced to spectate as the planet is destroyed for profit; forced to live in a thermometer that only rises and never falls, as the seasons are unlocked from each other, as the oceans curdle and swathes of land are consumed by fire.

But no-one would want to read such a book, I counter myself. It is too depressing and inevitable and as a species we do not have much stomach for the truth. But then, I counter-argue, others feel this way too, that we are on a sinister fairground ride, strapped in, dangling upside down on a rollercoaster controlled by people far more powerful and ruthless than us.

But I also want to write about time, about how I feel eternal, as well as an ordinary saggy fifty-year-old. I grew up in the 1970s, at least in calendar time (cue Chevrolet Impalas, avocado-coloured bathroom sets, 'Working My Way Back to You' by The Spinners on the radio) but in technological time we were living in 1730 or 1850 – any time before the invention of the combustion engine, basically. Around me the world has changed so radically in the time I have been alive it feels as if it has been torn apart. It is like being alive in a gassy inferno and a drawing made with invisible ink at once. What did I expect? I interrogate myself. Surely not stability, or certitude, or plenty? But I did. It never occurred to me that we would destroy the very basis of life, that I would witness this destruction in real time.

As I dry the dishes my thoughts are tugged in the direction of the Paleo People, those disciples of deep time. In 2009 I joined a scientific expedition to Greenland as writer-in-residence. I remember Anne Jennings, one of the paleoclimatologists on the science cruise.

'What are you going to write about?' Anne had asked, a reasonable question. There was no scepticism and disapproval in her voice, as there sometimes was with the scientists whose expeditions I'd joined over the years. 'What happens on the cruise, Greenland, time,' I'd answered. 'Which time?' she asked. There she had me.

I have mostly written fiction, a form which thrives on human energy, human emotion, human lifespans. If we want to expand the notion of the human we could say that fiction is an artform that rests on the fission of society. Its forms and genres are so varied as to be uncategorisable, but its storylines are often identikit: a woman marries into a family and has to assert her autonomy; a man journeys to a foreign land and acquires riches; three generations of a family struggle to maintain a farm; an orphan supersedes his origins and becomes a prophet; a group of survivors of a catastrophe build a new world.

Time is the clock that ticks through fiction, and landscape is often a background, not a generative force, or even a character. When I was offered the position on the Greenland cruise, I wanted to try to write about non-human time, and joining a paleoclimate expedition would make this task relatable. Even if I had already been there, to deep time, in Antarctica. I had seen how we humans were obliterated in its ivory pleats of forever.

In all, I would spend six years travelling to and writing about the polar regions, commuting back and forth like an Arctic tern. In Greenland I was hauled back in time on another hardship mission spent hunting the ghost of winter. But really there was another task I was reluctant to admit, even to myself. For research institutions, the value of writers was to be public communicators of science. But I was on the trail of something else. I would not know what I was looking for until I'd set foot on Greenland. There, the entity I was hunting for would reveal itself.

Baffin Bay, 69.2198°N, 51.0986°W

Uncharted Waters

August 2ⁿᵈ, 2009. Falmouth, UK. 17°C

Gyllingvase Beach is coated in fog. Droplets of rain drip from the razored leaves of chilly palm trees and the town beach is deserted. I've thought it so often in the British Isles: *if this is summer, I'd hate to see winter.*

After breakfast a van arrives to take us to the ship. Falmouth harbour is too small for many ships to come alongside, so they stand by on their dynamic positioning systems, stacked in the sea like planes approaching Heathrow.

We are sped out a few miles to sea via a pilot boat, an orange fibreglass lozenge that looks like a motorised Quality Street sweet. Outside the porthole windows we can't see our ship, or anything at all. But then we spot an outline in the fog, a stationary spaceship materialises: the *RRS James Clark Ross.* We hunch inside the pilot boat as spray flies over the deck. The *JCR's* captain comes on the VHF and tells us to approach. He has a British Airways pilot voice – lofty, posh, certain.

We climb a ladder and our bags are winched up. We are met with a wet deck lashed with warm rain. Two crews rotate on the vessel, doing four month-long stints. This crew is the 'opposite' to the one I'd encountered in my last voyage on the *JCR* in Antarctica, so I haven't met them before. I say hello to a tall, bald man, unusually thin for someone who lives on a ship, for reasons which will soon become apparent. This turns out to be Richard, the purser. He shows me to my cabin. They are numbered according to who is meant to inhabit them. Mine is Scientist 4. I am glad to be impersonating a scientist again; Scientist 4 is so much less adrift and suspicious than Writer 4.

On the bridge officers in black trousers and white shirts garlanded with gold epaulettes dash across its conference-room-blue carpet and pine furniture. Gruff seagulls stutter outside. Several cargo ships become visible in the mist, only a couple of kilometres away – the English Channel is one of the busiest shipping routes in the world. They are underway, their bows peel the ocean open. Forty minutes ago we were scraping the yolks from our plates in a Cornish B&B and now we are on a different planet, one of industry and precision. This is something I savour about attaching myself to such missions: the rigid vertigo we feel as the ordinary world falls away, as if it had never existed.

At last I am going to the Arctic, I think. I have always loved the word. The thrust, the bell-like peal of the concussive 'c's, the curve and flex of the *arc*, the nervous flinch in the *tic*. The word sounds like a place to pit yourself against, in order to discover the true nature of the world.

Greenland is the world's largest island (Antarctica and Australia are classed as continents). I have only ever seen it from the window of transatlantic flights, when I've stared and stared, riveted by its blank serenity. It will take us six days to get there. This does not seem very long to traverse almost the entire Atlantic Ocean. The ship travels at an average speed of about twelve knots, although at a push she can do sixteen.

'That's what you can write in your articles,' Alex, the ginger-haired third mate, says. Alex is, like all third mates, improbably young; he looks as if he has just escaped kindergarten. 'That we're bicycling to Greenland. It's the same speed as cycling.' Being on a ship did feel a lot like cycling, I knew, from previous experience – the circuitry of the waves, the grinding, iterative motion.

At the tea station Alex boils the kettle. The hierarchy on a ship is military-rigid, but all the officers, even the captain, take turns on the tea round.

'So,' he says. 'What exactly does a writer-in-residence do?'

'Write,' I say. 'Reside.'

'But what about us? Are we going to appear in what you write?'

'Maybe, maybe not.'

'If one of us were horrible to you, would you write about it?'

'Definitely. If you want to appear in print, just be horrible to the writer-in-residence.'

'Thanks,' he says. 'I'll keep that in mind.'

Alex shows me the manifest, a print-out of who and what we carry. On board are eleven scientists from the UK, US, Norway and Canada and twenty-five crew. The *JCR* will be not only a means of transport, but a life-support system, a society. Because we will not touch land until the ship arrives back in the UK in five weeks' time, we carry 1,236 cubic metres of Marine Gas Oil for fuel and 221 tonnes of fresh water (the ship is also able to 'make' forty tonnes per day from seawater using two freshwater flash evaporators). In the food stores are twenty kilograms of black pudding, forty large wedges of Red Leicester cheese, 192 beef burger portions, and ten boxes of iceberg lettuce. This is the one downside, apart from confinement, to ship life: being on a floating Wetherspoons pub, for weeks on end.

By this point I have been on ships only in the steely Southern Ocean. In the austral summer of 2005, the *JCR* transported me to Antarctica; the following southern hemisphere winter I'd come out of the continent on the last ship of the season, the *RRS Ernest Shackleton*. I'd expected to find ship life terrifying and confining and the entrapment among people with whom I had nothing in common to give me panic attacks. Instead I found I had a taste for ship life. On the *Shackleton*, when we encountered the Drake Passage in fifty knot gales, I was one of only two passengers standing while everyone else threw up in their cabins; in fact, I was so well the Bosun had taken one look at me working out on the Stairmaster as the ship corkscrewed through twenty-foot rollers and put me to work on the deck tallying cargo.

It is good to be at sea again, to feel the relentless revolution of the waves, and breathe the knife-sharp air. To go to sea on a polar research ship is to undertake a thrilling, potentially lethal mission. I allow myself to feel the lure of the unknown. This influence – it is more than a feeling, more like a companion, a chaperone – is so familiar to me, by now. It has driven me aboard this ship, just as it drove me out of the wrecked cars, cold trailers and hunting rifles of my childhood.

I had been to the Arctic before, spending the summer of 2007 on Spitsbergen in the Svalbard archipelago, but never at sea. What I did know was that the Arctic was the Antarctic's opposite in more ways than their names. Whereas Antarctica is a frozen continent surrounded by sea ice, the Arctic is a largely floating nation of pan-like sea ice anchored by one significant landmass - Greenland. The gassy ocean at the top of the world, stocked with bowhead whales and the unicorn of the sea, the narwhal, was the antithesis of the Antarctic's sink-like weight. The Antarctic was a dead zone, a ferocious planet moored at the bottom of the ocean. I expected the Arctic to be less totalising and more flexible than the Antarctic; it might permit us to know it.

On the ship we curve around the southern tip of Ireland. The light is already listing toward autumn. The sunset has a golden slant and the clouds are thin and high. The ocean beneath us is beguiling as always, a tilted blue that darkens quickly to black in the shadow of the ship. My eyes drink it in. In my experience, where we are going there are no blue oceans. In the polar regions the sea becomes a chessboard. Black, white, black, white. There, the world sharpens and gathers its power, like hoarding ammunition. There, a resoluteness lives in the land. I can't describe it – a spirit mentor, an aggrieved giant white bird, an old, old will. Once you brush up against this force, it is as if you are inhabited, switched into a different creature, no longer fully yourself.

Why is my bed moving? I bolt upright in my bunk at 4.30am. I remember I am on a ship in the middle of the Atlantic Ocean. I draw the blackout blinds to find a sky the colour of suet. The sea state has changed from last night. The waves outside my porthole are gunmetal and snarling. I know the seamen's term for this look from my previous journeys by ship in polar waters: *lumpy*.

At the lunch the captain, Jerry Burgan, joins us. It is he who owns the British Airways pilot voice. Social rank on a ship is as rigid as a monarchy. The protocol is that the ABs (Able Seamen) and officers must call him Captain, or by its synonym, the Master, or, colloquially, the Old Man. Scientists and uncategoriseables such as writers are allowed to call him by his first name. But the captains of polar research vessels are not regular mortals, so I stick to Captain.

As we plough through a lunch of sausage and mash, Captain Jerry tells us we are heading into uncharted waters. 'We have charts, they just aren't accurate,' Captain Jerry tells us, reassuringly. Earlier on the bridge I'd seen them whipping out huge flat paper charts that flapped like manta rays in the breeze, then poring over them with old-fashioned navigational dividers. The crew's usual habitat is the Antarctic; neither the captain nor any of the officers of the watch have been to western Greenland before.

We pass near the Reykjanes Ridge, the longest oblique spreading ridge in the world, and which separates the Eurasian and North American tectonic plates. The sea changes rhythm. The interval between the waves becomes longer; now they are long loping foothills. There is a feeling of overture to them, as if they are preparing us for the imminent arrival of mountains. Soon enough, we start shipping spray over the fo'c'sle, the term for when the foredeck disappears into the wave. Once in a while a 'gopher' – a large wave – douses the bridge, and the sky disappears into a wall of water. The queasy feeling so many people new to the sea report is not only seasickness, in my experience, but shock at the raw power of the dimension they have entered, its hostility, when they had expected beauty or leisure.

That night I can't sleep. It's not the motion of the ship that keeps me awake, but some other preoccupation I can't identify. I go up on to the bridge. It is overcast, moonless. At night on the bridges of ships, all light is dampened to cut reflection. You have to close the bridge door quickly, before too much light is let in, or the officers bark at you. I make my way across the space, bashing into some console or other, trying to see my way with only the dim light of the navigation instruments.

'I can't see,' I whisper.

'Give it three minutes,' says Alex, 'and your night vision will kick in. And you don't need to whisper.'

My eyes adjust. The fo'c'sle and foremast materialise out of the night, rising and falling against a graphite horizon. There is no moon tonight, but the sea has its own light source, foaming in phosphorescent arcs around the bow. I find it hard to believe we are really in the middle of the ocean at night, reliant entirely on the ship, on ourselves. I could write that it is like a film or a projection or a dream, but it is not any of those things. So much of my years trying to insert myself as a writer into places I don't belong will be about chasing this nameless feeling, which has something to do with witnessing the impossible made real.

Albedo

We file into the bar. On science cruises it is rare that everyone – scientists, Able Seamen, engineers and officers, cooks and stewards – are in one place together. We look around at each other and shake hands. Despite sharing a ninety-nine-metre-long ship, some of us have never met. We've come to hear the chief scientist and leader of the expedition, Colm O'Cofaigh from the University of Durham, give an illustrated talk on the aims of our cruise.

A slight, dark-haired man from Northern Ireland, Colm has a careful, methodical mien. The earth systems scientists I have met over the years are trained to be meticulous and this leaks into their manner: coolly evaluative, with carefully gauged opinions and emotions, even-tempered to the point of phlegmatic. When I first started out writing about climate change, I thought being surrounded by such steady characters might help me balance my more volatile nature, as if I could treat gusts of despair and restlessness like a chemistry or physics problem, inputs and outputs that could be measured and controlled.

Colm is illuminated by the weak gleam of PowerPoint. On it, graphs and charts sketch a story of lack: ice shelf retreats, leaking moulins – freshwater glacial depressions – helter-skelter ocean currents. On JR175, as our cruise is officially known, the goal is to study the behaviour of a major ice stream in Greenland in the Late Quaternary, between 500,000 and one million years ago. To do this we will run the ship over the seabed again and again, backing and forthing above where this ice stream once existed in order to map the seafloor using Swath bathymetry (on-board name: 'lawnmowing'). We will also radar the seabed of present-day ice streams, among them the fastest-flowing ice rivers in the world, which means, in climate change terms, the fastest melting.

Colm explains that eighty percent of Greenland is covered in ice – roughly 2.8 million cubic kilometres of it. The coastal fringe of the continent is rimmed by mountains; behind these, the ice cap rises to a plateau up to 3,600 metres above sea level. From it, streams drain the ice from the interior down to the sea, enticed by gravity, hydrostatic pressure and a process known as sublimation wherein the weight and friction of the ice exerts such pressure that the rock beneath heats up. As Colm speaks my notebook becomes peppered with glottal terms already familiar from my time in Antarctica: *retreat, scour, sublimation*, his accent sharpening their already precise edges. You can hear their meaning in their inflection. The relentlessness, the fume of campaign and loss and (rarely) gain sounds like overhearing generals discussing invasions.

One of the field sites of our study, the Jakobshavn Isbrae, a 520-kilometre-long tongue of moving ice, is thought to drain a monu-mental 6.5% of the ice on Greenland. It is commonly known as 'the fastest flowing ice stream in the world', so fast – advancing on average seven kilometres a year – you can see it moving, if you stand in front of it long enough. This is exactly what I intend to do when I eventually leave the ship in Ilulissat in a few weeks' time.

Ice streams are crucial to Greenland's ice mass and balance, Colm explains. Greenland is key in governing climate systems and especially the Gulf Stream; any change in ice balance will rapidly affect sea level rise, probably quite quickly, its effects probably hitting the British Isles first. The Isbrae's Formula One speed is not new. It is thought to have advanced and retreated very quickly in the recent past. Its last growth spurt was between 1650 and 1850 ('no-one knows for sure because there were no records,' Colm says, 'but our guess is around 1850.') But since then, it has mounted a steady retreat.

Colm shows us a now-familiar graph. Climate change science comes in two visual representations: vertical spikes or a steady subtraction, like an octopus furling its tentacles.

I glance around the room. We sit on the burgundy banquettes of the bar. Almost everyone's arms are folded across their chests, as if in protection. All faces, from the hefty engineers to the scientists to the electricians to the cooks, wear the same expression: clenched, like a box just closed. When you look at these graphs, it feels to me as if not only the quantity of ice but our whole world is shrinking from the hem in, reduced by a nameless force that is only partly about global heating. It is a willed diminution, a slackening of time in favour of a relentless spirit unleashed on the world: us.

In the half-light of the PowerPoint, I have a thought: all science cruises are novels. Always, in novels, there is an underlying question that forms a skeleton of the book – it is this structural feature, rather than length, that distinguishes a novel from a short story or a novella. Can one man rescue the truth in a totalitarian regime? Does colonialism produce psychopathy? Novels set out to explore these, knowing in advance they are answer-less, nudging their way through the darkness of ice and time. The question that galvanises our cruise is the same one that has taken root in my mind over the past four years. *What are we doing to the world?*

By 2009, the year of our expedition, the polar science fraternity was becoming accustomed to the idea that the two remaining great ice sheets of the world, Antarctica and Greenland, could conceivably collapse, possibly quite suddenly. Prior to the early 2000s, no-one, apart from two or three scientists who had been tracking climate change, and possibly, the oil company executives who knew from the 1970s that burning their black gold would warm the earth, had seriously thought that might happen. But the Arctic point of no return was judged to have passed in September 2007, two years before our journey, when the sea ice extent in the region was at its lowest in recorded history. The media was beginning to slough off its obsession with 'balance', i.e. cultivating a decreasing circle of nut-job climate change deniers, and to navigate the fact that climate change is not an event but a process.

In 2007, as the summer sea ice extent plummeted, for once the media had an event to cover. At that point the glacial past of Greenland and its coastal waters had been surprisingly little investigated, considering the role both will play in our future world. Climate change is a catastrophe for our species and the entire planet, but it has energised earth systems science, and especially the polar science community. Until that point, the climate change money and attention had gone to the Antarctic. The UK's Natural Environment Research Council (NERC) declared its interest in beefing up the UK's contribution to Arctic science. Our science cruise was one of these initiatives.

The briefing over, we all remain in the bar, galvanised by common purpose. This is one of the chief pleasures of joining science expeditions: as a writer you are on your own in a way so total it is difficult to describe, and which can so easily lead to a radical isolation very close to madness. To join a scientific expedition, especially one to a remote and dangerous place, is like being adopted into a raucous family. The madness becomes a collective purpose.

We had entered waters where we might see ice. The charts the officers showed me on the bridge said so: *Ice May Be Encountered Seasonally.* It was time to do our cold water safety training. We spent the rest of the day in apocalyptic cruise ship mode, doing sea survival drills, abandon ship drills, fire drills, acid spill drills.

On the way to some drill or other I ran into Anne. A ship's spatial dynamics turn us into medieval villagers. You can't pass someone without saying hello, which inevitably leads to a how-are-you, which often leads to an actual conversation. There was never any need for small talk on science cruises. You can just launch in with, 'What is happening to the albedo?' so I did. Anne was the person to ask, of anyone on the planet. She was an expert on paleoclimate and the second-in-command among the scientists on board, after Colm.

'The albedo of the Arctic is shrinking drastically,' Anne said. 'As soon as you lose sea ice, the albedo reduces from between eighty and fifty percent to ten percent. Even less.'

Albedo. The word sounds like a brass bell, also vaguely ecclesiastical. I had not heard this word much since my time in Antarctica. Albedo means *the proportion of light reflected back into space*. According to the *Oxford English Dictionary*, the word was used first in the mid-nineteenth century. It is derived from the classical Latin for 'whiteness'; in old English an *albe* was a white linen robe worn by priests. In ancient Greek, *alphos* was the name given to the disease known as white leprosy. There is a remnant of the root *alb* in the word albumen and in the old English term for swan, *elfet*, meaning white bird.

Later, I glanced out my cabin porthole to see we were chugging through fields of shattered ice. We had reached the hem of the summer ice field, the final remnants of icebergs now mere jagged pans of ice, transparent at their perimeters. Even so, they still reflected the light, helter-skelter like a field of smashed mirrors. White sabres of sun glinted from the ice floes. Suddenly I saw white swans, wings furled, their necks slowly being broken.

Now that we had entered the waters of Baffin Bay, terns and fulmars skimmed the surface, accompanied by their reflections. The sun turned black and oval. This was one of the visual tricks of the polar regions: the summer light is so fierce it produces chessboard reversals: black becomes white and vice versa. Mirages duly morphed on the horizon: a tractor, a miniature Moorish castle, a trampoline. These either turned out to be icebergs or delusions. On the radar we saw only seismic ships used in oil exploration or the Royal Arctic Line ferry which supplied many of the coastal communities in Greenland. I took to watching transatlantic planes streak through empty skies with binoculars. As we left the ordinary world, the planes looked like the hyper-advanced technology of an unencountered civilisation.

As we moved further north into Baffin Bay the sea calmed. In the Antarctic, such a becalming would indicate the presence of pack ice, either visible or over the horizon. But there was no more pack in Baffin Bay, not that summer. We came into a sea state I'd never seen in the Antarctic. The ocean was iridescent sheets of satin. All the Antarctic old hands on board could not believe we were in the middle of a 690,000 square kilometre arm of the North Atlantic Ocean. We went down to the aft deck to inspect this *Solaris*-like substance close up. The water parted as the ship glided through it, viscous, creamy, sighing with a voluptuous, unwater-y sound.

The foil was punctured by loafy bergs. More and more of them appeared as we edged north. They were much smaller than Antarctic bergs. But also they had a completely different character. Antarctic icebergs are monumental and stern, like continents forgotten by time returned to the planet to drift over the ocean, nursing their superiority. The Arctic version were casual sentinels, observably of this world, like ingénues out for a paddle in the ocean in flimsy kayaks. Antarctic icebergs look eternal, too big to risk dissolution. But actually, once in the ocean they melt quickly, within three or four months. They are frozen ships adrift in the ocean of time, practising for oblivion.

Paleo People

'What is the timeframe for Quartz-gold mineralization?'

'It's been dated to 1.77 to 1.80 billion years ago.'

'Which is when?'

'Late Paleoproterozoic, during the Ketilidian Orogeny, which began 1.85 billion years ago.'

'Good, Jean. You'll soon have your GCSE in geology.'

The Paleo People had assigned me the role of Science Cruise Dunce, the writer who knows a lot about nothing at all, and certainly nothing about what matters. It was good-natured, their aghastness at my lack of hard knowledge. They indulged and pitied me in equal measure. It was like being a cruise mascot. In any case I'd taken this part before and knew my lines.

Paleogeologists, geographers and climatologists all have in common not only the distant past, but several concepts of time: Real Time are the days, weeks, months that humans call time and which are usually elusive to measure in fossil records or other archives of the distant past. Human Time has a sliding scale: the 1.2 million years ago when *H. sapiens* evolved in landscapes like Lewa; the 400,000 years ago when they began to fan out from the incubators of the savannahs across other landmasses, right up until the 13-16,000 years ago when the Americas were settled by human beings for the first time.

But Deep Time is where the Paleo People really live. This can be anything from the Late Quaternary, around 2.8 million years ago, spiralling down into the mantle of time, the Hadean, the oldest of the four known geologic eons, when life on Earth did not yet exist, when the moon was an infant and the oceans had yet to condense and liquify. Mentally, Paleo People live on a planet still revelling in a fugue state, unaware of the alien species that will one day consume it. I envied them their world without us, only the long, slow dawn of time, arcing incrementally over a solitary planet, serene in its silence. My imagination now has a dual frame: the version of time which we are involved in, steadily eroding our common future, and the one in which we never evolve at all.

The next morning I draw the blackout blind in my cabin to find we are moving fast, passing iceberg after iceberg stationed at regular intervals in the ocean like fenceposts.

I dress quickly and go out on deck. I am braced for icy temperatures but the air is slouchy. The bridge thermometer agrees: it's twelve degrees in the middle of the Arctic Ocean. Two ABs are on the bridge with binoculars suctioned to their eyes, on the lookout for growlers – small, mostly submerged, icebergs. There is a slick of sun on the sea, turning it to zones of gold, bronze, magnesium. I think: this doesn't look like the polar regions. Where are we?

I slip down the stairs to the UIC lab (full name, Underway Instrument Control Laboratory). This is where we can access the internet on BAS PCs and our own laptops, plugged in with an old-fashioned ethernet cable. We are encouraged only to look at our email once daily, for two reasons: to not put pressure on the satellite connection's download speed, and to keep us focussed on our mission. I have not looked at my messages for three days. Not because of the download speed, but the ship is such a totalising environment it is difficult to believe any other exists.

J., my friend in London, writes that she admires me for putting myself in such an 'extreme male environment'. K., a friend who lives in Italy, writes, 'I don't know how you can do it. Six women and forty-two men. It must be like the Army.'

I look around, trying to see my surroundings through my friends' eyes: scientists, mostly but not exclusively male, hunched over their laptops, while in the corridor ABs – all men – flash by, on their way to winches and levers, their ears muffled against the sound of the below-decks in runway worker ear protectors. The officers of the watch are also men, although more and more women are coming into the profession and it is not uncommon to have all-female crews on the bridge.

I do a quick calculation: out of the twenty-five scientists and officers onboard, four of us (including me) are women. I do not feel outnumbered, or conscious of my gender. We are in a different realm, here. Normal rules do not apply. The ship is a rugged world, it is true, but I wouldn't characterise it as male. It is more that we are reduced to being merely human. I have always liked this about polar research, that you leave yourself behind, and also the world, when you step onto the ship/airplane/Ski-Doo. Ship life is often uncomfortable and sometimes, when the weather is bad, savage. But our mutual vulnerability blurs the distinctions of gender and class that would prevent us from understanding each other on land.

I write answers to my friends and go back to the bridge. Our destination was creeping into view. It appeared on the edge of our nautical charts. Disko Bugt, a wedge-shaped bay one third of the way up the long comma of Greenland's western coast. There we would start the survey of the face of the Isbrae. The charts tell me its real name, in Greenlandic: Sermeq Kujalleq.

The ocean is still a sheet of cellophane. All morning we slink through this sea-sky-mirror-world, the only sound the hum of our engines. It was as if the world has been stuffed with cotton wool. A fulmar flies exactly parallel to our starboard bridge wing and peers at us with its seed-shaped eyes.

After lunch, in the bar Anne shows us photographs of the Greenland landmass. The Isua sequence of sedimentary rock, of which the island-continent was part, was thought to be 3.8 billion years old, 'The area we will work is lithologically part of the Archean block,' she explains. 'The whole of the Hudson Bay area is still involved in what is called isostatic rebound; the land is still bouncing back, levitating upwards by over one metre each century, after having been pressed down by the weight of the Laurentide ice sheet in the last ice age.'

Isostatic; lithology; rebound; Laurentide; tectonic; plateau; geomorphology; precession. Geological language is like downing an ice-cold drink on a scorching day. We are alive on plateaux of time, undergirded by the patience of rock, it tells us. The land slides toward infinity while we humans calculate mortgage percentage rates and compare prices of carpet cleaners on Amazon. You can hear the hard glamour in it, the ordeal of just existing: through time, through rent and fire and cluster. To me it sounds not a language of the past at all, but of the distant future.

That evening we all ventured out on deck, lured by warmth. I took a cup of tea out to the fo'c'sle and for once it did not cool to an undrink-able temperature within a minute. In our shirtsleeves we watched bergs slide by.

The differences between the Arctic and the Antarctic accumu-lated. Arctic bergs were slack; they lacked the striated, compacted texture of many Antarctic icebergs, so dense and monumental they might as well be floating chunks of Carrara marble. And the mountains of coastal Greenland, while impressive, were less spookily grandiose than those of Antarctica. The beauty of the scene was muted, softer. The sea lacked the forbidding tar hue of the Southern Ocean with its brittle glitter, as if the ocean had been poured full of obsidian. The crochet of contrails left by transatlantic flights I spy on with my binoc-ulars were definitely absent in Antarctica. In the Arctic there were more birds, it was warmer, there was more *life*.

The ship clipped through the ice sculpture garden. On such a day, warm and so transparently clear it seemed we were cast in an eternal world of glass, cutlass light rebounding from ivory bergs, it was hard to imagine climate catastrophe – or any catastrophe, for that matter. The scene was so serene and commanding it silenced us. We stood with our cups of tea on the foredeck, a place of instant death when the sea was rough, sitting on a ledge with our legs braced against the Hawser rope winches, watching as we rippled past ship-sized icebergs.

There, I felt my first tug of anxiety about leaving the ship. Time seemed to reel itself in, like a fishing rod line that had suddenly snagged a bite. The beginning of our journey, only six days previously, the misty morning at the Falmouth B&B, seemed to belong to another dimension. Yet in ten days it would all be over for me, because I wanted to not only see Greenland, but to stand on its landmass. I wanted to compare it to the Antarctic: its smell, taste, its vibration. A day trip would be enough to take the necessary reading. But JR175 was ship-based only; no calls to Greenland were programmed, and for reasons to do with insurance and lack of pilotage, once off I was not allowed to re-join the ship. This was another way the ship was more than itself, or what it appeared to be. Like life, once you left, there was no way back.

Ilulissat

We sped away from the ship in the pilot boat, our wake cleaving the blue bay in two. From afar Ilulissat was an inviting sight, with its tutti-frutti wooden houses in russet, turquoise, lime green, all glinting in the high-octane Arctic sun. We were met on the wharf by the harbourmaster and his colleagues, men with jet black hair and square, ruddy faces wearing those glasses that darken depending how much light they are exposed to.

Our passports were stamped by a man in an Aran Isles-style jumper and fisherman's waders – the least-officious immigration officer I'd ever seen. He peered at me inscrutably from behind his glasses. I suppose he was wondering why the pages of my passport looked like those of someone in the British Army (ships, planes and helicopter insignias) or with a penchant for places no-one had ever heard of: Port Lockroy, Rothera, Mount Pleasant, Ascension Island.

Tim, Greg and I milled around, shocked at the warmth of the day – seventeen degrees. We had come off the *JCR* together. Greg had to get home for the birth of his first child while Tim had reached the end of his four-month-long tour at sea. The following day they would take a Dash 7 from Ilulissat to Air Greenland's main hub at Kangerlussuaq, then to Copenhagen, then home on easyJet – even if in Greenland it was hard to believe easyJet existed. I would follow an identical route in five days' time.

As BAS employees, Tim and Greg had the benefit of paid rooms at the town's tourist hotel. They invited me to share their taxi to the Hotel Arctic. From there I would walk to the Youth Hostel, the 'budget' option.

We pulled up in front of a sprawling low-rise with a giant insignia of a lantern-jawed explorer – Knud Rasmussen, probably – clutching a husky. I waited as they checked in, inspecting the reception area. A poster announced the dignitaries and personalities who had stayed at the hotel, a mélange of the saintly and the dubious: Ban Ki-moon, Jane Goodall, John Negroponte, Sepp Blatter. Beyond, in the bar, we could see Danish tourists sipping gin and tonics while on an outdoor terrace a barbecue was tended by tall white men wearing wraparound sunglasses. At the tables equally tall and blond people sat dressed in white and grey fleeces, watched by slavering huskies chained nearby.

I set out for town carrying my rucksack. The road took a steep descent to Ilulissat harbour, then rose precipitously on the other side. A bridge spanned a similarly vertical backwash to the harbour. In the ship's *Arctic Pilot* I'd read this was in part a buffer zone against the regular mini-tsunamis, called *Kanele*, which afflict Ilulissat, set off when icebergs in the bay capsize. As I walked I realised the smell of Greenland was familiar, a fume of rock and flint and no green. The smell of Antarctica.

I found the hostel. The whole structure shuddered as I walked down a corridor where doors led to rooms with no locks and four bunk beds. There was no-one at reception so I claimed a bunk. I had no idea if I would be sharing, or what I would eat, or what I would do with the four days I had planned to be in Ilulissat, which suddenly felt like a lifetime.

The silence was strange, chloroform. I remembered how long it had taken me to get used to the lack of sound in Antarctica. A feeling overtook me, a kind of dull panic. I had experienced this before, but not for a long time. I have always been prey to sudden gusts of emotion but I had learnt to control these over the years by blasting them with logic. *Why are you so afraid? You're in Greenland, there's no safer place on the planet unless you fear getting eaten by an off-duty sledge dog.*

To try to dispel the dankness that had overtaken me I sat down on my bunk to make notes in my travel diary. *Ilulissat*: the spelling kept tripping me up. I wanted to put in four 'l's and one 's'. In Greenlandic it means iceberg. I had once thought it a beautiful word, but now that I was there it was a mournful ring, something you would chant as an orison: *Ilulissat, Ilulissat, Ilulissat.*

Tim, Greg and I had arranged to go for a walk to see the mouth of the Jakobshaven Isbrae. I would meet them at the end of the boardwalk that stretched from the southern edge of town. On the way fat, floury bees barged into my mouth. The sun corkscrewed through my head in a familiar polar-regions way, as if it wanted to drill straight into your brain. The edges of the air were rouched with chill, but the sun scalded the skin.

At the ice stream's perimeter Tim and Greg were waiting for me, backs turned, looking at the view. As happens so often in the polar regions, the scene appears as a projection, a panorama filmed in Cinemascope, realer-than-real, then rebroadcast to your brain, which struggles to accept it. Because we cannot write ourselves into this image – it is too old, too epic – our brains refute its reality.

We saw a fjord clotted with a jumble of ice. In it, bergs sat quietly, like sleeping animals. We scrambled up to the top of a bald rock field coated with slippery, turmeric-coloured moss. From this vantage point we caught sight of a familiar silhouette: the ship, *our* ship, in the bay. It looked far away, and so small, surrounded by icebergs twenty stories high. Tim whistled. 'The bergs don't look that big from the ship.' We watched as the *JCR* glided around the corner of a berg and disappeared. I felt a tug, almost a pain, in my solar plexus. We turned back to the ice stream, drawn by its cold voltage. From there, we had ringside seats to something few human eyes ever see: an ice stream in motion, seen from the land.

I still have a photograph of that moment. I took it with a now-dead SLR propped up on a cairn with a remote timer. Like all photographs, it freezes time, consecrating it into an eternal present tense.

The three of us regard the ice stream. Behind us, it shifts and shuffles in a casual, almost lackadaisical way, destination-less. The ice stream is wedged between two brown-pated hills and shines like chromium hit by a sunbeam. The river of ice stretches inland, as far as we can see, a tumbling hummock of ice of all sizes and shapes: the wedges of future bergs, or ice shards as tall as apartment blocks.

At its mouth, the ice foams as it encounters warmer water. Vapour clouds veil the ice edge. Terns wheel across this disordered delta; when the sun catches them, they too look as if they are made of metal. Close to the stream's edge, the air is a degree or two colder. It singes our faces. There is a restlessness in the air, like an electric current.

Now that we are on land, we can witness what from the sea had seemed an inscrutable, almost invisible, process. We see that *ice stream* is a misnomer for what is actually a renting, raging vein of white blood coursing from the hydrostatic pressure of a beating heart. Thirty-five billion tons of ice are pushed out of the ice cap every year. It is the Greenland landmass, bleeding.

In the photograph the *JCR* is just visible in the background, perched on the surface of Baffin Bay, backlit by an amber sun. Tim, Greg and I wear an identical expression: slack-jawed with awe, also abandoned, also triumphant. In my years in the polar regions such expressions would flare like matches. Briefly the pathos of our situation freezes into place. We mourn, or are afraid. The world is dissolving, the world is on fire. And yet it is so very beautiful, and we are so privileged to see it. This is the tumble of moral injury and the reversed sublime in which we live now, all the time. That moment in front of the ice stream was one of the few when this circus of emotions coordinated itself inside me, so that I could be a witness, at least. Yet I can't grasp its meaning for very long. The present rushes in to fill me, and the feeling melts away.

Hotel Arctic

The next morning I wake early. My throat has a metallic taste, as if I have swallowed spoons in my sleep. I drink two litres of water. The hostel is still empty, although someone had tried to enter my bunk-bed room at some point in the non-night, maybe two or three in the morning. Half asleep I had stood and remonstrated with them. The sight of a half-dressed woman trying to stand up without falling over in a sleeping bag must have scared them off.

I set out for my interview. I would meet with the director of the town's museum. I had arranged this by email while on the ship. He did not respond initially, and I'd had to chase him a week after my initial message. Eventually, he'd responded in a half-hearted way. *Come if you have to*, had been the tone.

The town of Ilulissat reveals itself to be hundreds of broken *Centre Pompidou*s. Everything that is normally inside is outside: household piping clad in foamy insulation, strollers, tins of snowmobile oil, half-dismantled sledges. Huskies with café-au-lait eyes are chained to their railings.

After fifteen minutes I have the town's measure: there is the Zion Church, the Museum dedicated to the life of local-born explorer Knud Rasmussen, three cafés, a white weatherboard bunker that turns out to be the Bank of Greenland, one supermarket, Pisiffik, where the chicken, fish and beef are frozen Danish brands and the only coffee is Nescafé grounds.

What is not the town is rock covered in moss and the occasional cleft of grass and the flowers of the Arctic and their heartbreaking names: dwarfwillow, crowberry, *Tripe-de-roche*, juniper – its tangy smell slicing through the vacuum-chamber odourlessness of the place – Arctic cottongrass, dandelion, saxifrage. As I walk it occurs to me these names are not that: tags, labels. They are actually the language of the land. Inside these words is a truth beyond emotional truth but related to it. Whatever it is, it has been established a long time ago, before the land dared to believe we would exist.

I walked into a russet-coloured, rare two-storey building. There was no reception or guide to greet visitors. Within a moment I found myself shaking hands with a sturdy man dressed in a fetching woolly jumper. This was the director, I was to understand. He led me down a wide corridor, the floorboards creaking under our tread.

His office was lined with sepia portraits of Greenlandic men in furred parkas. I entered and settled into a chair facing his desk.

Then I had a strange experience, the first of several I would have in my remaining days in Ilulissat, a period that would be shorter than I expected. It was something like déjà vu, but also not. The moment in which the director would turn and sit at his desk, his atmosphere of weary familiarity. The moment when he would tell me about the town's main industries, fishing, hunting, and tourism. The gazes of the men in parkas on me, their serene command. The smell of frozen resin emanating from – what? The frostbitten wood, which had been hammered into place in 1923 – all this was rehearsed, or no, that's not right, it was pre-real. An echo settled in my head. Whatever either of us said, I heard it two, three times, each time passing through a deeper chamber of rebound. The final echo boomed, a voice I hadn't heard before. It was bass, convincing. It came from outside me.

I realised the director was waiting for me to speak. I tried to shake off the voice, the way that time suddenly clung to me in clumps.

I managed to put my questions to him. In halting, aspirated English, the director told me the town's main industries were fishing, hunting and tourism. The latter was growing as fast as Baffin Bay's climate warmed, he said. In a few years, up to 50,000 people were forecast to visit Ilulissat annually, many of them motivated to see climate change up close. 'A form of disaster tourism, isn't that what it is called?' he said, his voice flat. People came to see the ice stream, they went to the museum, he said. Sometimes they went out on boat trips into Baffin Bay.

He tilted himself forward and made a cupping gesture with his hands. 'You know, winter is like a bowl. The bowl used to have depth – deep cold, for months on end. Now it has been flattened to a plate. The bowl is gone and may never reform.'

I looked at his hands, which he held in front of my chest, like a priest offering communion. His skin was the deep russet of many Greenlanders. But the edges of his fingers were frosted, with fissures of pale skin. He saw me staring. 'Frostnip.'

I knew its effects. I had suffered it too, as a child, and later while taking care of a barn-full of twenty-five horses in temperatures of minus thirty-five for several winters. I still had the tell-tale mottled capillaries on my thighs.

I composed myself. 'I've thought this too – the land is changing to accommodate what we are doing. As if it is flattening,' I said. 'Maybe it is trying to forget that it ever knew winter.'

The director gave me his first and only look of curiosity. His eyes, I noticed, were flint, neither green nor grey nor brown but an amalgam, like malachite.

'What do you think will happen to Greenland?'

He shook his head. For a while I thought he might not answer. Then he leaned forward in his chair, quick, as if suddenly jolted by electricity.

'People put that question to me all the time. It is cast as a question but what people want is a statement of loss. There will be many opportunities. The mineral and oil explorers are already here. I think Greenlandic people will adapt, over time. It took perhaps five thousand years for indigenous people in the Arctic to adapt to cold. They will have to adapt to heat much faster. Who knows, maybe in a thousand years Ilulissat will be some kind of mecca for people in parts of the planet that have become too hot to survive. It will be Paris!' At last, he smiled for the first time in our conversation, although there was a rapier tilt to it, and it extinguished itself as soon as it existed.

I walked back to the hostel. Each house was surrounded by a moat of Arctic off-season detritus: rusting Ski-Doos, Nansen sledges. This, and especially the restless dogs, attested to the fact that winter was the real season in Greenland. Everything was waiting for it. The town looked exposed in summer, like someone caught in public without their clothes.

Four teenage girls approached on the other side of the road, dressed in black shorts and singlets despite the fact it was only twelve or thirteen degrees, Doc Martens on their feet, sunglasses on, thick glossy hair tied into braids, an army of little Lara Crofts. The girls did not even flick an eyelash in my direction as I pass. Why should they? I interrogated myself. You don't live in their community, you may as well be just another of the North Face-clad tourists in pristine hiking boots who chew through imported beef at the Hotel Arctic. *You are a ghost here.*

At first, I think I have made that familiar transition, from internalising my self-hood as an 'I' perspective, then pivoting to the you, the accusatory second person who is also me. But I realise there is another voice, cached within this familiar me-you. It has an aspect of the external about it, the fractured sound of someone speaking to you but from very far away.

I stop. The town is silent. All noise seems to be suctioned out of the air in Ilulissat. Maybe the thickness of the ice absorbs it, like giant baffles. I feel a stern gaze on me. The gaze has something to do with the voice hidden inside my own. It knows I am a ghost to this land, to the girls, to myself. It knows I am unconnected to the souls that had roamed its long white tongue all these sleeping epochs. It knows I can't read the ice, I don't know how to feed the dogs, I am a mediocre kayaker, I am a useless envoy from the realm of Tube trains and Aperol Spritzes, of twilights translated effortlessly to night.

The voice begins to speak to others – who, I don't know. Some sort of audience. It speaks about me in the third person, as if I am not there. *She needs the succour of trees. Look, her mouth is dry.* Yes, I am more thirsty than I have been in my life. I drink another litre of water but it is like drinking air. I say to the voice: *I'm not the ingénue you think I am. Cut me some slack.* But it only laughs, a hollow, grandiose sound, albeit less mean than human laughter. Is this the same voice I heard as a child, growing up on the shores of a giant inland ocean on a glacial island not so very far away? It sounds neither male nor female, human nor animal; it lacks the booming authority of film trailer voiceovers, my standard internalised voice-of-God. I'm not completely sure I am not generating it myself.

I should be more worried by this concert of voices, but then I am a writer. It could be that all writers have multiple voices circulating inside them all the time, like an internal village. I know writers who say their characters speak to them. ('I even went on holiday with my protagonist once,' R., a novelist friend, tells me. 'He was good company but kept wanting to drink mojitos in grim bars. It was terrible but I had to do as he said.')

The voice I hear in Greenland is beyond such whimsy. It has something to do with an external story, one I am probably not equipped to write. In fact, the voice is the narrator, for once, not me. There is one story this voice is writing, over and over: we are deleting ice from the world; we are pummelling the land; we think the planet was constructed just for us. The land is truculent, like a dog out on a walk that suddenly clutches its haunches and refuses to move. But soon, I feel, another phase in its rebellion is coming. This is my real task as a writer: not to write about the privations of ship life and how writers in chips-are down situations are absurd with their delicate feelings and their lack of command of facts, but to record the stages of the land's grief and outrage.

This internal voice may not be about Greenland or Ilulissat, I realise. Rather I hear it because I have been at sea for nearly two weeks, and now I am on land. Here, I feel exposed to a force I cannot name. I miss the ship and its regimented, reassuring rhythms. I miss the things we saw in Baffin Bay and which were indescribable. I have exiled myself into a hardscrabble town that, as much as I try to resist the comparison, reminds me of where I grew up, a place not too far away, as the crow flies. But beyond that I can't say why I feel so desolate in this pretty-sounding town which I had expected to like. We sometimes say, *I felt so at sea*, meaning so confused and disoriented. But for some reason the sea has become land for me, over years of attaching myself to ship-bound expeditions, and the land the locus of my unease.

Greg and Tim leave the next day. I watch as they are bundled into a taxi with all their duffel bags branded with *British Antarctic Survey, JR175*. We say hearty goodbyes. The polar fraternity dissolves the usual English reserve and we hug each other through our down-filled vests. I will exchange friendly emails with them for a few days or weeks, then I will never see or hear from them again.

I walk back into town to the hostel where I write up my notes from the interview with the museum director. I am trying to follow through on my research plan, but another story has begun to etch itself into existence, like a bathymetry image of a long-unseen ocean floor.

The next morning I wake in my sleeping bag after a sleepless night. The all-night howling of 1,500 sled dogs has not helped. But the real reason I haven't slept is the stern voice I have heard since setting foot in Greenland has hectored me all night, badgering me with answerless questions. *What are you doing here? You're not a journalist, you're a novelist, which means a fantasist. Why did you leave the ship? Everything worth writing about was there, on the ship, not here.*

My throat is dry. I sit up in my bunk and look out the square hostel window. Brightly coloured buildings that look like they've come out of an IKEA packet dot the horizon. Beyond them, granite boulders knead the sky.

I don't know where the desolation inside me has come from. It feels unfamiliar. If I have to describe it, I could say it is not personal, to me, but about the whole project of being human, which feels unbearably hollow, suddenly, a grim masquerade of doing-and-knowing.

I make the decision in the way I do when I have reached the end of my ability to accept my limitations: unthinkingly, without resolve. I pack my belongings, leave the hostel, and walk to the Hotel Arctic. At the reception I ask for a room for two nights and lay down my credit card like a sacrificial animal. I find myself walking through the hushed, darkened corridors of the hotel, opening the door on a *2001: A Space Odyssey* room, all curved chairs and bulbous lamps. A double bed beckons. I lay down and finally go to sleep.

I changed my flight to leave the following day. It cost me hundreds of pounds. Waiting out the rest of my self-exile, I sat in my hotel room and drank ten-dollar beers, the curtains shut against the glare of the ice field. Flung off the ship and out of motion, on land with the silver-haired tourists and their slashes of lipstick over thinning mouths, I mourned the things I would have seen had I stayed on the ship and which were indescribable. I would probably never be anywhere as beautiful as Disko Bay again, with its roseate icebergs prowling their way to dissolution.

I would not follow through on my research plan. I would not get the story. I had left the ship too soon, and now I was leaving Ilulissat like a fugitive, I was running back to the feathery trees and mild rain of the mid-latitude summer. I was always doing this, arriving too late and leaving too early, afraid of overstaying my welcome: in people's homes, their hearts, their countries.

In ten years' time we will learn that the warming of the planet will inevitably melt the Arctic sea ice and the Arctic Ocean will soon be ice-free in the summer. The last time the top of the world lacked its ice cap was 2.6 million years ago, when humans existed only on the sere plains of east Africa and were about to fashion stone tools. By around 2040, when my life will likely have ended or be ending, Ilulissat may be as much as eight degrees warmer than it is now. As the museum director speculated, it may become a place people go to in order to be cool for a change. The ice stream will still be there, prodding, nudging, coursing its way into the dark heart of the ocean. I am a privileged witness to the last interglacial period the planet may ever know. Because of this never-ever game, the cold nodes of the world will always provoke in me a see-saw of an empty ecstasy and a grating desolation.

It is my last night in Ilulissat. Evening sidles up to the Hotel Arctic like a gigantic ocean liner. The sun has not neared the horizon in two months, but tonight it will strafe the sea. The dining room is bathed in its cold glow. The chained dogs orbit their small wooden doghouses, restive. Blackflies coagulate in clouds. I go to stand on the granite shield where the hotel is perched on patches of greasy mustard-coloured lichen. A gunmetal strip of cloud is laid across the sky. The sea is steel, serene.

At midnight the sun approaches the horizon, fizzing and dimming like a hurricane lamp turned low. The sun glares red on the flanks of the icebergs. Tourists turn their flint eyes on the spectacle in unsmiling silence. The huskies rise as one and howl as it sinks into the ocean.

PART III

THE END OF DESIRE

March 26th, 2022. London, UK. 9°C

I am shocked to discover I have Coronavirus. After two years of pandemic, after taking eight international trips even at the height of so-called lockdown, I thought I must be immune, or that something was protecting me from it. A bit like death, I thought it only happened to other people.

On the other hand, I'm worn down, so it makes sense. I've been working too hard lately: Ukraine benefits, rail replacement services, emergency trips to my job in Norwich, two hundred kilometres from my flat in Stoke Newington, writing emails until midnight, being on the receiving end of thousands of university edicts: *Reminder from PGR office. Mandatory Finance Management Course – Action Required. Total Compliance.*

I feel I am going in circles. I can't see how to change anything. The country is in the grip of a right-wing government that seems as if it will go on forever. I have lost my European citizenship or, more correctly, it has been stolen from me and the result is I feel as if someone has taken a cleaver to me and sliced me neatly down the middle. I knew, of course, that reality could be reprogrammed around me by malevolent forces. I've read *Nineteen Eighty-Four*, Arendt, Kafka: all the totalitarian classics. But until the Brexit vote and then the pandemic I have never been subject to this furniture rearranging exercise.

The virus feels similar, as if it has been mustered to enact the same charade. Also, it is strange to have a virus you know you have never encountered before inside you. I can feel its unfamiliarity, the way it rummages inside me, knocking on doors that neither the virus nor I know existed, or whether they will open. But I don't yet feel so ill, so I wander around my flat, trying to make up projects to get me through the coming days. Maybe I could rearrange my haphazard bookshelves?

Not for the first time, I note how many books I have by writers who are masters at rendering landscape: Farley Mowat, Tim Winton, Margaret Atwood, Barry Lopez, Damon Galgut, Annie Proulx. I realise I haven't read Proulx's short story, *Brokeback Mountain*, for years. One evening I battle through the subscription offers and paywalls on the *New Yorker* website and find its original appearance in 1997 there, intact.

The story is not a classic in construction, with its epic sweep, its swift collapse of years and even decades into a couple of lines here and there. In fact is structured more like an artifact from early twentieth century literature. It is not naturalistic but heavily narrated; it does not linger on any one moment as a turning point. This time I read it not through the prism of character, but landscape:

> Dawn came glassy-orange, stained from below by a gelatinous band of pale green. The sooty bulk of the mountain paled slowly until it was the same color as the smoke from Ennis's breakfast fire. The cold air sweetened, banded pebbles and crumbs of soil cast sudden pencil-long shadows, and the rearing lodgepole pines below them massed in slabs of somber malachite.

I love how the mountain pales *slowly*, the sense that behind the paling there is a watcher – not only Ennis, but the narrator, or God, and how the phrase 'slabs of somber malachite' is stratigraphic, geological, indicative of the remote voice of the narrator being broadcast from far away. Even the American attempt to erase any trace of Romance languages from English by reversing the 're' at the end of Latinate words has an effect: *somber* is so much more sombre than *sombre*.

On Brokeback Mountain, things turn, as they do in stories, for Jack and Ennis.

In 1983, twenty years into their relationship, Jack and Ennis go packhorsing into a less promising-sounding range, the Hail Strew

River drainage. Proulx's language turns on a dime. 'They left it to wind through a slashy cut, leading the horses through brittle branch wood, Jack lifting his head in the heated noon to take the air scented with resinous lodgepole, the dry needle duff and hot rock, bitter juniper crushed beneath the horses' hooves.' We hear the burn of the wind in Proulx's phraseology – dry needle duff, bitter juniper – as well as the hardening of the edges of Jack and Ennis' predicament.

I hadn't thought it the first time I'd read the story, but now I understand that Jack and Ennis' love affair is not only with each other but also the place; their love is only possible *because* of the place. In this sense landscape and desire are the same thing. They can hide in plain sight amongst them, two cowboys, disposable rural labour of their era, derided and under-paid, forced to fund their lives in back-breaking rodeos or hurling frozen hay at cattle in winter. Moving from the early 1960s and the 1980s, the characters stand on the cusp of this new era of awareness of climate change and the Anthropocene, but as characters they essentially belong to the past. Remnants of the American west and the taciturn, folded-in people these empty mountains and plains sculpted, Jack and Ennis are envoys from an age so recent we can still smell it, when nature still appeared largely unaffected by the human.

I am a product of that age myself. I come from a province whose glacial wounds are still so visible you want to suture them with words. Nova Scotia sits tilted into the Atlantic Ocean, the cranium of its body the lunkheaded island where I grew up, Cape Breton. I lived in Nova Scotia until I was thirteen, and since then I have found its mirror in unlikely places: South Africa's Western Cape, the lonely coast of southern New South Wales in Australia. But it was in the Falkland Islands, flung out from the tapered tip of South America, where I found its true twin.

Like many of the places I have spent time, I remember the Falklands by their emotional atmosphere rather than anything that happened. For me they were a site of ecstasy and failure, entwined. The

Islands' landscape was not that different from Montana or Wyoming, in some ways, as was their cargo of sheep farmers bent against the wind as they battled through clumpy tussock grass to reach muddy Defenders at the end of the world.

I close the *New Yorker* screen, leaving *Brokeback Mountain* to rotate inside me for another decade or however long it will take me to read it again. Meanwhile the story reignites the question I've left on smoulder these past three years, since I have begun thinking about writing a book about landscape and language: if the land could speak, what would it say, and how would it sound? If the Anthropocene holds that human beings have become a geological agent, it could be that we are ourselves becoming geological. Do we enter the strata, the rock-fable of the planet; do we become a part of a metaphysical machine? And if so, might we finally be able to hear it, the voice of the land?

East Falkland, 51.7963°S, 59.5236°W

Las Islas

The flight is only an hour from Punta Arenas, but it feels longer. We endure the usual rollercoaster ascent after take-off as the plane, an LAN-Chile workhorse A320, battles the cut of crosswind. The Beagle Channel recedes. Soon we are treated to a frieze of the Southern Ocean, with its cobalt-and-white horses below, roiling like a pot of boiling water.

In this windy annexe of the planet, there is windshear even at twenty-four thousand feet where the air is normally smooth. The plane veers through the sky like a drunk dragonfly. No-one on the flight touches their beer because they're too busy clutching both of their arm rests and/or reciting Hail Marys. Cue the inevitably burly man in the row behind, probably a contract oil or construction worker from the UK on his way to the Islands for hardship pay, who pipes up. 'I never thought I'd be glad to see the Falklands, but Christ, bring them on.'

For me, this journey was the last chapter in an unlikely itinerary that had begun a week before in Cape Town, South Africa, where I was living. I'd flown with Malaysian Airlines across an unnervingly divertless (no airport in St Helena then) Southern Ocean, nine hours on a lonely transect to Buenos Aires. Because the Argentine government forbids any direct air links with the islands, I had to hopscotch down the length of Chile, flying first to Santiago then to Punta Arenas.

I'd been to the Falklands twice before, on the way into and out of Antarctica, so I know our first glimpse of the islands won't offer the nervous flier in the row behind me any succour. There they are on the horizon, a cluster of jagged shapes, taloned like the feet of an eagle taking hold of the ocean, mustard-coloured, with edges the loamy colour of black pudding. The sea below us becomes consumed by more ragged blotches with spiked protuberances, like virus molecules. The Falklands archipelago is actually made up of some 770 islands. They are shattered sparks flung out four hundred kilometres from Tierra del Fuego, the terminus of the Americas.

Planes do not land in the Falkland Islands as much as fall out of the sky. The land rises up to meet us. Beneath us we see no roads, no sun-glint on roofs or other signs of civilisation, only stern grey rivers that are too big to be rivers tentacling across the molecules. I grab the armrest out of intuition and sure enough, another crosswind swats the plane. When the plane's wheels snatch at the runway the fuselage is still flying sideways. The pilot wrenches the nose straight and I feel the push of the rudder in my solar plexus. Yellow hills stream past in the window before halting their angry rush.

'Welcome to the Falklands/Las Malvinas.' The stewardess' voice curdles over the PA – maybe she has also feared for her life. The plane erupts, not into flames, but ecstatic cheers and applause. That's how you tell the Falkland islanders from everyone else on the flight: they always clap on touchdown.

Landing at the military airport, optimistically named Mount Pleasant, I felt a stab of fear and a trilling, lonely dread of nothing in particular, other than emptiness, far-awayness, of having somehow ventured over the lip of the known world and ended up in a lair of elaborate sea monsters and rapier light.

The light. How to describe it? It really does feel as if you are being stabbed in the eyeballs. It slices sideways, tobacco-coloured, drunk on wind. The light is not only itself here, but part of a three-dimensional reality called *lightwindsky*. You feel it immediately, the savage scrutiny of this dimension. It is there even in the way the sea behaves; waves do not come ashore but claw at the edge of the land. The southern Atlantic can wear different colours: teal, petrol, cobalt, sapphire, jade, tar. In the Falklands it is the colour of cinders. These ashy waves collapse on the shores only to be driven back into the ocean by the wind, among the strongest in the world. I have never seen this anywhere else – an ocean doubling back on itself, repelled and chastised.

I arrived in the Islands on Good Friday. Stanley, the 'capital', never hopping on the busiest of days, was deserted, apart from the Upland geese that congregated on the green between Ross Road and the Sound and who regarded me neutrally as I passed. I ventured out to the West Store to get groceries and found it closed, as were all the gift shops that slung stuffed penguins at the cruise ship tourists. There was nowhere else I knew of to buy food, so dinner was a muesli bar brought 12,755 kilometres away at Whole Foods in Stoke Newington.

I simply walked into the house where I would stay – keys and locks are unnecessary in the Falklands. No-one steals because it's impossible, or at least very difficult, to do anything, let alone leave, incognito. For one, there is the matter of one of the largest British Army bases in the world down the road, stocked with under-occupied squaddies all too ready to pursue anyone who tries to nick a TV.

The house where I was staying belonged to Veronica Fowler, an English schoolteacher who had lived in the Islands for decades. I had first met Veronica two years previously on my way to Antarctica. She had agreed to host my residency at the High School. She even gave me her house, moving out of her bungalow for the six weeks I would be there to stay with her partner – an outpost generosity I would come to know was typical of Veronica.

Her house was old, by the standard of the Islands, with a commanding view of Stanley Sound. The paint on its weatherboarded exterior was peeling under the onslaught of wind and salt. The living room was a kind of personal museum of the Islands, decorated with nautical charts and drawings of the many ships wrecked on the reefs and promontories of the archipelago.

As I walked around the streets that Easter weekend in search of chicken thighs or tinned tuna or rice – any food, essentially – I tried to bat away a leaden panic. What was I doing here? Unlike when I'd been in the Islands before, in 2005 and 2006, when I'd had the protection of a large organisation, the British Antarctic Survey, this time I was on my own, although supported by grants from Arts Council England and the Shackleton Scholarship Fund. I had signed up to teach creative writing at the local high school and to a group of adults as an evening class. I also pledged to write a short story set in the Islands. Even if I knew from experience that declaring in advance you would write on this or that specific subject was bait for the writing Gods who dished out inspiration and failure in unequal measure.

On Easter Sunday Veronica appeared. At seventy she was a still-beautiful woman, a bit Julie Christie with blonde hair, blue eyes and breadknife mouth. She brought me a casserole, which I immediately devoured. 'The West Store was shut,' I explained, but the truth is I was always starving in the Islands. 'Ah,' Veronica said, with weary familiarity. 'That happens a lot. It's the wind.'

Veronica and her husband John had arrived from the UK in the 1970s, she told me, 'on the milk boat from Montevideo.' Her house was stocked with photographs of her and John when they were younger. I'd already inspected these and arrived at my verdict: a handsome couple, objectively too good-looking to end up in such a hardscrabble place.

She and John never expected their marriage to break up after thiry-five years, Veronica said, in her gravelly voice. 'But that's just what happened.' I thought this one of the most honest summations I'd ever heard, and direct – a characteristic of the islanders, as I would come to know. The phrase resonated; things we least anticipate end up happening, and you would never believe it, should someone tell you how it would all turn out in advance. The capacity for life to generate surprises is one of the many reasons why we should never know the future.

On her way out, Veronica stopped in the kitchen. 'My best friend was killed just where you are standing. It was the first day of the war and an Argentine shell landed on the kitchen. We'd been having a cup of tea. Half the kitchen was destroyed. But don't worry,' she said as she went out the door. 'It's not haunted.'

Stone Runs

The high school was the biggest civilian building in the Falklands. Located at the western end of Stanley, near where Ross Road petered out into boggy hills, it looked eerily similar to the regional high school I'd attended in New Brunswick in eastern Canada, and which could have starred in a teenage blockbuster titled High School Hell. Lockers, bells, bloodcurdling screeches and sudden lunges by clumps of teenagers, the mashed pea smell of the cafeteria – how had I survived this in the first place, and why had I willingly come back?

I was chaperoned to my classroom by an amiable teacher. The children I'd be working with were fourteen to sixteen year olds. It seemed this age group were largely occupied in terrorising the English teacher, just arrived from New Zealand, staking out her apartment at night and throwing rocks against the window. The only students who responded with vague interest in reading and writing were recent incomers, quiet girls from St Helena. These girls read from their exercises in class in abashed whispers. 'Writing is not a task, it's a way of being present in the world, present *to* the world,' I told the class as I exhorted them to keep a diary of things that happened in their lives. To this the kids scoffed, 'Miss, nothing ever happens here.'

The adult writing workshop took place in the evenings in the offices of the local newspaper, the *Penguin News*. Among my students was a retired mercenary who had been in the Biafran Air Force. He was textbook ex-gun for hire in his Army issue green jumper and iron handshake. Apart from writing about birds of prey he had completed an exhaustive study of all aircraft downed by both sides of the Falklands War and wrote novels in which the word 'Mossad' appeared on the first page. 'No-one's ever taught creative writing here before,' a soon-to-be-retired policewoman in my adult writing group affirmed. 'There were two women who came from UK to teach crafting and knitting, and the shearing men who come from New Zealand to teach the Chileans how to do it.'

No-one had written about the Islands in any literary form, as far as I could tell. My students told me there were no songs or customs that were not imported from either Scotland or Argentina. In the town archives I read accounts of the early days of settlement, beginning in the late 1700s and through the 1800s: 'A remote settlement at the fag end of the world', recorded one of the Islands' early Governors, in 1886; a sad salvo from a Spanish priest summed it up: 'I tarry in this miserable desert'.

Charles Darwin, when he visited the Islands on two separate occasions in 1833 and 1834, was struck by a geological feature. The islands are draped with distinctive, unique rivers of stone. Huge boulders unravel down the hills in streams, like stony hedgerows. Darwin was confounded. How could these have formed? The Falklands had never been glaciated, as far as anyone knew. Darwin posited they had been 'shaken into place by earthquakes'. His hypothesis was correct, although the duration was off. The earthquakes had happened over millions of years, from frost action and thermal fatigue. The stone runs were a remnant of the freeze-thaw cycles of the last ice age.

The exilic air of the islands was geological, as it turned out. Their true home is far away from their present location. The Falklands were first welded to what would become southern Africa; projections of Gondwana show them wedged between contemporary Durban and Antarctica. They then twirled around to become part of Antarctica, then pivoted again 180 degrees to become aligned to the south American continent, as if torn between divorcing parents. The rocks in Lafonia, in the south of East Falkland, are the same as those in the Karoo, a semi-desert terrain of southern Africa – which meant that when I'd flown to the Islands from Cape Town via Argentina and

Chile, I'd travelled twenty-three thousand kilometres to walk on the same rock.

The next day Veronica came to visit again, knocking politely on her own front door before bursting into the kitchen on a blare of Falklands wind.

We sat in her living room with its commanding view of Stanley sound, clutching mugs of tea, watching yellow squalls blow between the hills at precise intervals, as if generated by a machine.

'Until ten days ago the weather was perfect,' she said. 'Autumn comes fast, here. One day you are tending your petunias or whatever and the next there's snow on Mount Tumbledown.'

I told her that the volatility of the weather in the Islands had struck a chord in me. 'Cape Breton is so similar to the Falklands, but so different at once. It's like I'm looking at one landscape and seeing a cracked mirror image of the other.'

This was true, but a crude description of what I felt. It would take me weeks to understand what the two islands were to each other, at least in my imagination.

'It takes a long time to get used to the Islands,' she said. 'But when you do, you don't want to leave.'

'Do you ever go back?' I asked.

'Back where?'

I realised my error. There was no *back*. Britain, if it ever had been home to her, was now a distant, foreign country. 'I can't bear what Britain has become,' she said.

'And what is that?'

'Grey, harassed by money, worried about survival. You can't drop by for dinner, there is no generosity, everyone is counting their pennies, calculating whether they can get by with serving only one avocado at dinner instead of two. Here people are well-off, because apart from food there is nothing to spend your money on.'

She travelled, Veronica told me, she had great friends, she still enjoyed her job teaching English. Satellite television delivered the world to her living room. 'It's not like the old days when one ship came a month and it took two weeks to get anywhere.'

For a certain type of person, I could see that the Falklands were a refuge from the vicissitudes of the world. But the world had come to find even this outpost, a long time ago, although in the Islands everyone seemed poised for another invasion, possibly tomorrow. They practically had their bags packed.

My eye snagged on Veronica's books – an entire wall of them, sensibly arranged.

She caught this. 'Ah, that's The War Shelf.' Our eyes ranged along them. *Sea Combat off the Falklands; Weapons of the Falklands Conflict; Don't Cry for Me, Sergeant-Major; Iron Britannia; The Fight for the 'Malvinas'; Beyond Endurance; The Falklands War; My Falklands Days*, by the former governor of the Islands during the invasion.

'That's the thing about wars,' I said. 'They're never really over.'

She hummed what might have been an agreement. 'You know, it's strange,' Veronica said. 'I remember you as tall and dark-skinned.'

'Yes I know, I remember you as much taller and blonder.'

'I *was* taller and blonder!' She gave her sawdust laugh. 'And younger.'

Memory and time collude to distort the internal pictures we carry of each other, we agreed. As for the Islands, they were also mutating, layering their present selves over the memory I had carried with me since I had last seen them in April 2006, on the way home from Antarctica. Then, twenty-three of us came out of the Antarctic on the threshold of winter. We were still galvanised by the chromium cult of the continent. Just leaving Antarctica, especially across seas as rough as the Drake Passage, felt like a feat, an escape from exile. I had carried the ecstasy of survival and projected it onto the Islands. Then, Stanley seemed like Paris: ten types of shampoo! Coffee with real milk! Mobile phone signal! We Antarcticans went everywhere in ragged, boisterous gaggles. We had not been alone for months or years, and the idea of sitting on your own in a café or waiting by yourself at a bus stop felt daunting.

'Did I ever tell you about my friends Bev and John?' Veronica interrupted my reverie. 'They moved here from Grimsby thirty years ago because they figured it would be so safe – no crime, the nuclear fallout would miss the entire area. And the next thing they knew they were POWs in an Argentine military camp.' She fell into a silence. When she looked at me, the spark of mirth in her eyes was extinguished. 'You can never tell the future, you ought not to be able to.' It sounded like a warning.

The silence of the house was ruffled by the sound of the trees scratching against its walls. We went out onto the deck that overlooked the Sound and stood in the still-mild air, our hair whipped by wind. Around the edge of the deck Veronica had planted a row of shrubs whose branches looked like pipe cleaners. 'Wind break,' she said, ruefully. 'Nice try.'

'All the vegetation here is unfamiliar,' I said. 'But also familiar. In fact, everything here reminds me of Nova Scotia, but it's slightly off. Even the metal roofs make me feel like I'm home, but a very displaced version of it.'

Veronica said, 'A lot of people say that the Islands feel like other places but also totally unique.'

Many of the photographs I would see in the Islands' archives, the library or in people's personal albums I would see looked so much like those of my family in Cape Breton: hard-eyed parents with large broods, organised, resolute, completely self-sufficient. They died of exposure or appendicitis, just as my forebears had.

But they were not the same at all as the people of the island I came from, eleven thousand kilometres from Port Stanley as the cormorant flies. The Falkland Islanders were hale, withholding, with brisk surnames – Biggs, Peck, McKay – instead of the hard-drinking, flamboyant Scots who had brought me up. And the landscape was so different: much harsher, denuded of trees, tallow bogs, the strangeness of the seasons – snow in August, buttercups in November. They were nothing like each other, Cape Breton Island and these seven hundred untidy scraps in the southern Ocean. Then why did I feel so at home?

Something about the Falklands reassured me, I realised. They were bleak and remote, but they had a Spartan resolve. There, I did not feel so harassed by the things I had not done, the success that had eluded me, by being forty and on my own. The Islands were stuffed with misfits and characters. But I would never belong. I knew, from being an islander myself, that no matter how long I spent on them, islands were closed societies: you belonged from the beginning, or never.

I don't want to write any more about departures and arrivals, I thought. I only want to arrive new in the world, without a past. I don't want there to always have to be a story. I am tired of antici-pating events. I don't want to know about what I felt before, I thought. I wanted to come to the Islands because in a way I cannot define I have always been here.

Veronica left to return to her partner's house. Alone with the non-ghost, the silence of the islands pressed itself on me. Wind scraped across the roof. The Araucaria trees, the only species that could with-stand the exposure of the Falklands, whispered against the house, pinned there by the wind.

Although *wind* is a flimsy word for the force that circulated in the Islands. It felt more like slabs of air. The power of the prevailing westerlies was partly because the wind travelled all the way from Australia without encountering any land to break it. Also, the winds of the Southern Ocean tend to be katabatic, meaning gravity-driven, rather than the standard version created from the tension between sea-atmosphere moisture flows. In the Falklands, eighteen miles per hour was considered a calm day. When the wind was really bad, you couldn't walk or even stand.

The hills north of Stanley gleamed as if lit by an internal electricity. Some days they were ashy, others dull gold, like the flanks of a palomino horse. The Islanders told me the light was so bright because there was no land for thousands of miles in all directions, if you discount Argentina, that is, which they tended to do. (Their reference points were Australia and New Zealand, not Antarctica or South America. They even had something of an austral twang in their accent; they crush their 'e's, just as Kiwis do.) Yet I could not find the words for the light, either its quality or its effect. I sat next to Veronica's fridge, rearranging her standard English teacher's fridge magnet words into something that might approximate it: *Amber and steel baubles swinging on pendulums. Iridescent gasoline. Light swipes the air like the paw of a giant tawny animal.*

Mostly I sat in Veronica's living room and looked at the sepia photographs of the many ships that had been wrecked in the islands, listening to the silence. Apart from the wind and the fighter jets that tore the skies open every day on sorties from Mount Pleasant, there were no sounds, or rather any sound that had the temerity to approach the island was instantly bundled away and drugged by the wind. Within this silence my own voice boomed in my head. I would need to find something to drown it out.

Storm Petrel

I rummage in my files on a Macintosh laptop so old I'm convinced it will not even turn on. It is grey, the thickness of five or six MacBook Airs, and so heavy I can't believe I used to cart it around everywhere with me in a bulky laptop bag which now also has the mien of an ancient artefact.

The Apple icon lurches to life, fizzing out of a field of grey. A previous version of me appears: forgotten folders, nameless novels I abandoned before they could even grow fingers or toes. Whenever I go looking for stories or even novels I've given up on, I feel like a mineral prospector. I might hit paydirt, or just dirt. Mostly is a reason you didn't follow through on a novel or story, but still these discoveries elicit a hollow, uncanny feeling, like unearthing a corpse.

Here it is, in a folder I have named 'Sur' – Spanish for south. (All my folders are titled in this way, with trick names, as if I expect MI5 to come looking for them one day.) The file bears the story's name: 'The End of Desire.'

A sign, painted on driftwood, declared the house's name: *Desire*. She pushed the door open.

Hello?

The house answered back: *who are you?*

She found a note on the kitchen table, written in a torturous script: *Gone to Town. A.* The foreman was supposed to be here to meet her, to show her what to do.

She felt something brush against her calf. She looked down to see a cat, as orange as the Cheshire Cat but whippet-thin, entwining itself at her ankles. Two green eyes looked straight into hers with the unmistakable glint of hunger.

She went to the cupboard. Two packets of Bisto and a tin of Horlicks leered at her. *What is this, the 1950s?*

She found she'd actually said this out loud. Already talking to herself/the cat: not a good sign.

In the drive two Defenders were parked, nose to tail, like reposing camels. She found the keys in the ignition of the first and turned it over, expecting to hear the tell-tale cough of an empty petrol tank. But the vehicle hummed to life.

The cat had followed her. She scooped him up and put him in the passenger seat, where she found what looked like a disembowelled jumper covered in orange fur. The cat settled into it.

'Ok, so you're used to being driven around,' she said, just to hear the sound of her own voice. The only other noise was the wind, which had a strange insistent note, like a piano key permanently depressed.

She found a piece of rope where the seatbelt was supposed to be. In another moment she understood the reason for it: the driver's door didn't close properly, but also had to be tied with another piece of thick green rope, wound round the handle for that purpose.

Now trussed into the car, she bounced down the driveway, past the electricity windmill, tanks full of foul-smelling liquid, solar panels, All Terrain Vehicles, storerooms full of gnarled bits of halters and stirrups, past the blades, now rusted, of what looked like a small combine harvester.

The wild yellow sky followed them. The road was gravel and so bleached it looked like mercury. The sky was partially overcast, but when the sun nudged between the clouds swords drove themselves through her eyes.

'What do you think?' she asked the cat. 'Why is the light at 52 degrees south so different from 52 degrees north?'

The cat closed its eyes sagely.

'Ok,' she shouted, over the wind. 'You're not much of a talker.'

The premise of the story is well worn: a woman, a Londoner all her life and recently turned forty, inherits her aunt's farm in the Falklands and flies there to sell it. But something (I hadn't yet decided what it would be when I started) detains her, and she stays. I wanted to write about exposure, physical and emotional: what it means to leave your ordinary life behind, with all its routines and comforts, and to start again at the end of the world. The character doesn't like the islands at all, but something in the place beseeches her to stay. She heeds this desire that is not hers, but at a cost.

It is quite possible that this fictional woman has been living an existence very like mine, paying six quid for papayas in Whole Foods in a rapidly gentrifying area of London, writing book after book as if her life depends upon it – which it does, or at least the electricity bill. This woman has never entertained a nostalgic quest for a place that never was, the place most people call 'home'. Yet, in being lured to the Falkland Islands out of guilt and duty, she finds that the place she has been living in and in her weaker moments calls home is not that at all, but an imposter.

The first thing I notice is that the prose is bouncier than my usual grave wave-roll, more of a semi-comic tone. What was it in the Islands that encouraged me to write this way?

In the Islands, my character finds something necessary in their shear and tear. I remember the wildlife missions I undertook there: the penguin colony at Volunteer Point, the plains of Lafonia, to West

Falkland to stay on a farm. Everywhere I went I saw corpses: eyeless dead rams, dead seal pups staring from those blackest empty sockets, dead petrel chicks, all their eyes plucked out by the Striated caracara, an aggressive native falcon the islanders called Johnny Rooks. The storm petrels that rode the bucking waves, along with mollymawks, skuas and pampa teal, all flying ellipses above olive-green clumps of kelp that looked like giant beached octopuses. On the shore, hunchbacked Rockhopper penguins scurried with their jam-red eyes and yellow hairpieces. And me, scrambling over boggy hills of tussock grass and Diddle-dee, thinking again about the light. I didn't like it, but I was in love with it.

'Tell it slant', the memoir-writing manuals opine. Meaning it is less interesting to render reality head-on, with forensic sincerity, than to come at the truth of a situation like a sidewinder, that snake that propels itself laterally across Sahara sand. We do not live our lives within the geometry of plot, but at an oblique angle, trying to reassure ourselves that something as outlandish as reality is actually to be expected, normalised, accepted as real. Defamiliarisation sounds like a psychiatric disorder, but this word signals one of the principal pleasures of reading (and writing). It refers to the way that a writer can conjure up the ordinary and familiar in a completely new way, and so present the world as the vivid enigma it was when we were children.

The Falklands was a slant place. Everything was either recognisable but weirdly skewed, or completely outlandish and weirdly skewed. Where *were* we, exactly? Like distant Guernseys or Bermudas, Britain shipped car parts and Waitrose own-brand marmalade across the seas. But on television we received *CNN en Español* from Argentina, even though we could not travel there direct from the islands, while the British Armed forces channel broadcast back-to-back brainless programmes about retiring on Gran Canaria. Prices in the West Store supermarket were in pounds and pence but they sold *Vital Agua*

Mineral, Leche Colun and *Huevos Frutagro* brought by ship from Chile and Uruguay.

We seemed to be in Latin America, but a colonised version from the nineteenth century, tethered as we were to the distant mothership of the British Isles. Meanwhile the human world had a fume of Aberdeen or an Alberta Tar Sands town. Oil and gas prospecting money had drawn check-shirt wearing men from rigs off Libya or in the North Sea to the islands to blow up the seabed of the Patagonian Abyssal Plain. In one of two of Stanley's brasseries these men dined with women in white bandeau dresses who picked at £20 plates of Chilean scallops. Drunk squaddies collapsed face-down on pool tables. No-one said hello or asked your name. Women, especially, eyed up other women as if they'd just stepped into a cage fight. Veronica's husky voice echoed through me: 'forget making friends here, except for other foreigners.' 'I'm not a *foreigner*,' I said. 'I'm British. And you've made lifelong friends, you told me as much.' 'That was before there was any money here,' Veronica replied.

Outpost larceny was nothing new, in the Islands. In the 1800s, the many ships wrecked on their shores were looted by the 'kelpers', as the islanders were then known, almost before the last bedraggled survivor set foot on the islands' stony shores. The original salvage merchants, the kelpers had made their living and sometimes riches from others' misfortune.

Now it was nature's turn, in the form of recently discovered oil and gas fields, ransacked in the service of Range Rovers and Air Max trainers. Lucre had come to the islands at long last, and in the bars and restaurants in town – there were no theatres, cinemas or other public spaces – I practised my habitual eavesdropping. 'I heard Mike got the contract at the South Georgia field. The lawyers in London told him to form a shell company. We're thinking Grand Cayman.' A strange energy circulated. It almost had a smell: coppery, not too unlike gun oil, come to think of it. Greed. What was I doing in this Klondike place with its petrol skies?

The story I discovered on my old laptop continues for a few more pages, before petering out into the white desert of fictional possibilities. 'The End of Desire' – I don't know how the title came to me. What does it mean to reach the end of desire, and how do we know when we have arrived there? Surely to reach the end of desire is to reach the end of life.

I abandoned it because in the Falklands I could not drag out of me that molten energy, akin to obsession, necessary to write. The story slipped away from me and went into permanent orbit, an orphan. Yet its premise is instructive. Like me, my character recoils from the island's punitive isolation and lack of 'culture', but is compelled by everything else about them: by the blue fracture of Falkland Sound; how the islands appear as scraps of pennants ripped in the wind; by the emptiness of the Southern Ocean and its relentless circulation; by the stone rivers that cascade down their turmeric hills. She finds a bleak thrill in the cold, bell-like peal of the names of outlying islands of the archipelago: Speedwell, Sea Lion, Steeple Jason, Bleaker and Carcass. The fissure and fury of the islands runs through her, too. It is possible that she has to stay, in order that the islands can cauterise an inner wound she did not know she had sustained. Maybe the islands are only the living embodiment of a shattered internal landscape.

A mysterious tractor beam keeps her there, an Orphic quest for an unmet Eurydice. She basks in the islands' strobe-lit Hades, high on the coconut smell of the gorse in their spring, on the hunt for a tendril of a former self. She feels her fate is bound up with them, but this is a delusion. The Falklands are just another staging point, a place where she can test her capacity for destitution.

I spent six weeks in the Falklands all in all that year, watching autumn turn to winter and snow dust Mount Tumbledown. In the end I failed to write my story – the only time I have ever reneged on an official commitment. So much of the task of being a writer is about sheer follow-through. I don't like unfinished business. Stories don't like it either. They hold their abandonment against you and haunt you more efficiently than the non-ghost in Veronica's house.

In his final book, *Embrace Fearlessly the Burning World*, Barry Lopez tries to diagnose his love for the land. If you love it enough, he suggests, the land will love you back, and even heal you. The land may be subjugated to our will, turned into an inert resource, but Lopez writes of it as a sentient presence, 'still present, vibrating in the shadow lines', beneath the asphalt and concrete. He goes so far as to suggest an intimacy with place can be almost erotic. The land can ease 'a particular kind of longing' that results from 'intense, amorous contact with the Earth'.

But passion for the land is not the same as passion for a human. The fervent love of a place on the planet, a landscape, lacks the dark magma core of love for another person. So much has to be aligned, in order to write fiction. The inner and outer landscapes have to be ready to speak to each other. During those years I kept driving myself back to the ends of continents, shuttling between unforgiving capes that had seen the foundering of talented navigators, looking for that congruence. But possibly I am making too much of all this (how unlike a writer) and those end-of-the-world places, a role which Falklands probably plays the best of all of them, are the sorts of places someone like me should never set foot.

Yet they are inside me, playing on repeat. I am flying in and out to 'Camp', as Islanders call anything that isn't Stanley – from *campo*, countryside in Spanish – in the tough little Britten-Norman Islanders, the pilots so calm as we flipped into hurricane force winds. Turquoise coves, the grass sheath of the land, lying golden in bays and coves of a hard blue. The marshy twilights as we flew back into town, platinum beaches punctuated by penguins. The bunched, secretive Araucaria trees, like gnarled wizards. Black-browed albatross slicing across waves.

And now, more than ten years after the last time I stood on their yellow, sentinel mountains, in so many places – Kirkenes, on the border between Arctic Norway and Arctic Russia, in the grassland interior seas of northern Kenya, in fishing ports in Nova Scotia, in north Norfolk on a stormy day, on the marshy plains of Rio Grande do Sul in Brazil, even when I am running in a particularly stiff wind at home in London. Anywhere sere and jerry-built with a note of wind-rushed loneliness and I think: *the Falklands!*

PART IV

BITTER PASTORAL

April 28th, 2022. London, UK. 10°C

A cold spring. At the end of April I return from Namibia to London. Now, when I land in London I no longer say to myself, *I am home*, but *I am back*. Britain has not felt like home since the 2016 Brexit referendum. I have two passports but no home. In fact, if anywhere qualifies as home, it is southern Africa.

It is spring and London is almost translucently green. In Namibia, the southern hemisphere was tipping into autumn. The *koppies* of the city were burnished by a slanting sun and the arid air was like inhaling paper. We are not evolved to move around so quickly. The body still thinks we are where we have left. The sense of place rotates, dislocated, inside the body. Another possibility: our bodies are that instrument geologists use, the clinometer, which measures incline, or the magnetic compass that detects direction. They carry place within us, not just its memory. We call it jetlag, but it is landscape-lag; world-lag, even.

My body has imbibed the sombre ordeal of Namibia: the gigantic puff adder smashed on the highway to Hosea Kutako airport, the amber slant of autumn light, the acacia islets in a sea of copper, hundreds and hundreds of kilometres of it, the teenage schoolgirls wearing white socks and Mary Janes, the lean black men who wait, carrying used paint rollers and dressed in blue overalls in the carpark of Megabuild on the outskirts of Windhoek. *Paint your house, ma'am. Paint your house.*

I am reading Daisy Hildyard's long essay *The Second Body*. 'What do an American barn owl, a Zimbabwean Hippopotamus and a Norwegian Reindeer have to do with you? What they have in common is that they all have a relationship to your body – they are all, in some sense, your responsibility.' Hildyard's argument is that we forget we are bodies because we live directed by abstract concepts (like money) and ignore the body, largely, until something goes wrong.

Bodies do have their own lives. You never know where they will end up. Bodies are uncanny, in the context of climate change. They have a second life, as Hildyard argues, as destroyers, entities out there doing damage, almost unbeknownst to us. The body that takes a flight from Windhoek to Addis Ababa and then to London, burning seven thousand tonnes of carbon in the proceedings. The body that consumes Brazilian-flown papaya in the mornings. The body that turns on the computer or the oven and has no choice but to be tangled in a web of electricity provision that leads to an oilfield in Venezuela or Saudi Arabia. Once considered discrete, the boundaries between bodies are being dissolved, Hildyard writes, by climate change. We are all implicated in each other's bodies and our bodies are melting the planet we live on. Perhaps we are all becoming one entity.

Meanwhile the weird Catherine wheel of where-I-am-not rotates inside me. The thin men and their geometry of inequality, the guillotined snake, the heat that builds in the day like the insides of a kiln. All this will also dissolve, soon. The body will incorporate it. Membranes, fluids, adipose memories. But my addiction to the motion that is destroying us will persist.

This trip was my fourth to Namibia. What keeps drawing me back? Certainly not its immigration policy. Its border officials begrudge your entry to the country, stamping your passport for only a week longer than you intend to stay instead of the three months you are legally entitled to, quizzing Belgian families about their intended hotel and then ringing the hotel, presumably to confirm that the family or hotel or both exist or if the blond-haired family already wearing bush hats even though they are in an airport are not part of a Russian money-laundering conspiracy.

I know why they do this. They are zealous of their country's natural assets, its copper, uranium, its relatively high standard of living, by African measures, and want to make sure everyone knows. But the real lucre is Namibia's watermelon sunsets and tallow deserts, the Crayola Crayon colours of the orange-red spectrum: Burnt Umber; Brick Red; English Vermilion; Venetian Red; Mango Tango. Also its emptiness. Three hundred kilometres is considered a short trip by road and if you see more than twenty cars in one place you know there has been a funeral or wedding. Namibia is the second least densely populated country in the world, after Mongolia, and its immigration officers are certainly committed to keeping it that way.

My first trip to Namibia was in 2010; I returned the following year and again in 2013. Something drew me back. At the time I didn't know how to explain it. Often as a writer you are working on instinct. My intuition about Namibia would lead me, eventually, to write a novel inspired by it but not set in the country itself, titled *Fire on the Mountain*.

'Fiction has to be set somewhere'. I say this, sometimes, to my students, to emphasise the importance of space and place in establishing fictional worlds. Very often this is interpreted as being an identifiable country, but place is much more subjective than that. My thesis is we don't fully understand place, either in lived experience or in imaginative writing. That is why it is so hard to evoke well in literature. Place is not a setting or backdrop but a zone of interest, a collusion between the body and atmosphere. It is also a consciousness. Each novel I write sounds as if it was written by a different author, and that is because the voice changes, depending on the frequency of its setting, of the landscape. Sometimes I think the land is the narrator and I am only its amanuensis.

In Namibia I found a place that was the embodiment of an inner landscape I didn't know I harboured. In this place was another hidden entity, an old foe, a dual creature, part man, part snake. Surely there must be a mythological equivalent, like the Centaur, but I don't know what it is. It was also a district of grief, where time moved not forwards but backwards, transporting me to a moment of previous departure I could not forestall.

NamibRand, 25.10317°S, 15.958586°E

The Land With No Fat

September 20ᵗʰ, 2011. Damaraland, Namibia. 35°C

The drive from Uis to Hentiesbaai is poles, road, poles, road, a melon sun. In the truck we are chastened into the silence that grows, as the moment of departure draws closer, between people who will soon never see each other again.

It is September, early spring in the southern hemisphere, and we pass under the mantle of coastal fog that extends ten kilometres inland, a product of the refrigerated Benguela current that flows all the way from Antarctica. The fog intermingles with sandstorms, turning noon into sepia twilight. Just visible in the murk are glossy ribbons of kelp, fur seals, the hunched shaggy forms of the beachcomber hyenas who eat the fur seals, fishing shacks, knackered bakkies and hulking remnants of ex-ships.

The Ovambo called this place 'The Land God Made in Anger'. The Bushman called it *Bitterpits*. I read this word first in Stephen's poems when he was alive and all was well and thought: that sounds like the place grief lives. I should go. Maybe I thought I would open a door at the edge of its vanilla deserts and find him there, waiting, his gas-flame eyes and wife and two children standing behind him, his future intact.

Several things have changed since my last visit to Namibia: Stephen is dead. He died suddenly in April this year of stomach cancer. In February I spoke with him on the phone in Cape Town; he was so weak he could not have visitors. I never saw him again after that day in St James when we climbed the mountain and he told me that I would come back to teach with him and I thought, *no*.

I met Stephen Watson in 2010. We were colleagues who did not have enough time to become friends. Stephen invited me to teach at the University of Cape Town, where I held a fellowship for a year, while living between London and South Africa.

Stephen was a mountaineer, a walker, a marathon runner, a swimmer and like many men and women who live in the climate and landscape of the Cape, physically perfect. A prolific poet and essayist, he'd published what many people considered his best collection, *Return of the Moon*, in 1991. These poems were based on the Bleek/Lloyd translation of the San (Bushman) language and lore, now stored in the archives at the University of Cape Town.

Lately, Stephen had published an essay about the Cederberg, 'Bitter Pastoral'. In it he calls it 'the land with no fat'. He was one of the few writers I've read who was able to put on the page the charisma of a place, to document the quality certain places have to pump oxygen into your heart.

In his essay on the Cederberg he writes about the power of places to compel the mind and the imagination, quoting Plato, who, as Stephen wrote, 'once spoke of place itself as "a veritable matrix of energies"'. D. H. Lawrence called it 'spirit', or *genus loci*. Smells, the light, the currents of air, the sound of water running over stones – he was able to invoke all the senses in the struggle to convey something of the mystique of place. The pastoral has always been an oasis, Stephen wrote, a place of retreat and succour. The Cederberg was Stephen's bitter pastoral, the place he felt most attached to, which bound and released his imagination at once, as many English writers revere the wet hills of Devon or the withholding shiver of the fens.

The day before I left for Namibia the previous year, Stephen took me for a walk on Table Mountain, up the trail that snaked from behind his house in St James. We reached the top and he turned to me and said, 'You see how easy it is to become obsessed with this place.'

We were looking out onto False Bay, into the bony fingers of the Cape Folds, the three-hundred-million-year-old mountains that unfurl in front of the Southern Ocean. It was a Sunday in April, early autumn on the Cape, but a thin heat remained. By then I knew well how the Cape created obsessives, slaves to the thrill and salve of places of such optic splendour you could hardly believe your eyes. Europe lacks the dimension to produce such awe. It has been too trammelled. The Western Cape, like Namibia, is one of those places you could still imagine before the arrival of *H. sapiens*. It had retained its independence of spirit, its indifference to the human.

We surveyed the view. 'You'll come back next year, and we'll do this again,' Stephen said. I had to stop myself from saying, *I won't. Something will go wrong. I don't know what, but it won't be like that.* The certainty was there, automatic, pre-scripted, I didn't need to think of it at all. I knew, I think. Not what would happen, but what would not happen. I wonder if this counts as oracular knowledge. A premonition.

Six months after my trip to the Cederberg it is September 2011 and Stephen is dead. I decide to undertake a walking expedition 150 kilometres across the Namib desert in Namibia, from the interior of Damaraland to the Skeleton Coast.

A charity, Rhino Alert, has organised the walk to raise funds for its important work safeguarding the Namib desert's population of black rhino, now a critically endangered species. I love walking. I want to be a flâneur, to 'abandon myself to the crowd', as Walter Benjamin wrote. Although I don't know it at that moment, I will soon train to be a walking safari guide. This will give me an excuse to walk and walk and never halt until something (an elephant, a mamba, a gorge?) stops me. Motion has become the same as emotion, or better said one is an evasion of the other.

But the real reason I am doing the walk is in a kind of homage to Stephen and his work, specifically *Return of the Moon*, his collection inspired by an archive of Khoi San poems, stories and drawings which together make up the most comprehensive picture of the pre-invasion culture of Bushman lives and world-view. Before coming to Namibia I spent time at the University of Cape Town, reading some of the original texts in translation as well as Stephen's poems. I first saw this landscape in the words of its original inhabitants, then through Stephen's re-re-rendering. By walking across the ancestral terrain of the hunter-gatherers of the Namib and the Kalahari deserts, I have the impression I will also traverse the thinking process behind Stephen's poems. Perhaps, I think, I can perform a writerly resurrection.

I had been to Namibia before the walk, to Damaraland and the Kaokoveld in the north. I found almost unsurvivably hot, a place where people, Europeans and their descendants especially, wither instantly and if you do not know how to find water you die within two days. At the heart of Damaraland is Namibia's own Uluru, a mountain called the Brandberg (Fire-Mountain) that rises drastic and out of nowhere to command an aortal view of the land.

What was the connection between me and this country on the outer edge of Africa? It had none of the lushness of the tropics, which I have always been drawn to. It was a country of taxidermy ranches and camping stores and fading Bismarckian mansions (some built with snow roofs, despite being located smack on the Tropic of Capricorn – those Germans were taking no chances) that looked as if they'd been teletransported from fin-de-siècle Munich. Meanwhile giant bronze dunes sidled up to its country towns, stocked with oddities such as horned adders with eyebrow scales that make them look like enraged elder statesmen, and the gemsbok (oryx) which have their own air-conditioning system in their nasal cavities which allows their brains to keep from exploding in temperatures of forty-five degrees.

In the beginning everything is possible; in the middle one or two outcomes are likely; the end is inevitable. So said Aristotle, the first theorist of narrative. But if we reverse this chronology we start with the inevitable and end with the possible, we see more clearly what has been discarded or lost along the way. We glimpse the ghost of luck, map the path not taken. Why should it be this way? we ask, retreating along the trail, as if we have seen a lion. And everyone in Africa knows this is what you do when faced with a lion: stand your ground, then back slowly away.

The Skeleton Coast

Day 7

The last day. We file through rocky cuts that will lead us down into the riverbed. It is forty degrees. The sun stings our arms. The red walls of the canyon close in on me. We pass lone eland and oryx standing sentinel in sand rivers with those cool nodes in their heads preventing their brains from exploding.

I try to dispel the impression that this walk has not happened before, exactly, but that it is happening in reverse-time. Everything that happens has already taken place. We are walking through not a scripted future, but an eternal past.

We are near now to the Skeleton Coast. We feel its coolness, feel the presence of the wrecks that bleat like scars, the *Eduard Bohlen*, the *Otavi*, the *Dunedin Star,* beached spacecraft lost in the fog Angolans call *cassimbo*. The Bushman call this place The Land God Made in Anger. Anger can have a velocity, even a beauty.

Suddenly our feet are walking on sand. 'Ah, the Ugab river,' Jan sighs.

That night we make a fire in the middle of the Ugab, or where the river would be if there were water. Here the dark is so total that if we do not sit within a metre of the fire we disappear. Our fires burn bright at the beginning, then wither.

For these last six days we've walked twenty to thirty kilometres every day through dune grass swaying in the wind. It looks not at all like land but a reversed sea. I see only us, the horizon, Jan's dog bending into the land.

Eight or nine kilometres beyond the Rhino Alert camp the finish line awaits. This is such a trifling distance for us now that Jan doesn't even put on his trainers. His feet are swathed in bandages, as are mine. Alice, the Frenchwoman who does logistics for Médecins Sans Frontières (so not a delicate creature) has a centimetre-wide hole in her head from when the metal peg of our spinnaker-shaped shade tarpaulin hit her in a freak accident, and the tip of Helen's toe will fall off within the next two days. We've even had to bandage the dogs' paws; Tiki, the little herding dog, pads along on Band-Aids.

This week we have all turned brown and lean like kindling. Jan has gone beyond tanned; his face, caramel at first, is rubber. Two tourmaline eyes stare out from it. He looks like one of those hard men in *Grand Theft Auto*. Around him the air is electrified, unhappy. He is alert, taut, but there is something of the same surrender of the language of this place in him, too: *succulent*, *ephemeral*. Such voluptuous words for a thorny place. Like the buffalo thorns that attach themselves to us, driving an inch down into our flesh, we absorb them until they are dissolved into our bodies.

Day 6

Before dawn we rise and stand by the fire, shaking off mist, dew, scorpions. Waking up after surviving another tentless night among hyena, jackal, elephant and leopard feels like coming unpeeled into the world. The spring sun rises by six. We watch Orion fade with the night westward, into the Atlantic.

We walk all day. What do we think about while we walk? For once I don't think. The wind roams through me.

My life now is blisters, zinc smear of sunscreen, trying not to sit under the tick bush nor step on a puff adder, migrant fears that waft in and out as I try to take the measure of the emptiness of this place.

Scar of bustards in the sky. The whisper of the swishing land-sea-grass in the wind.

This area of the Namib was formed when the Atlantic retreated three hundred million years ago. Jan picks up petrified ostrich eggs we find smashed on the ground. The ground is covered with the shattered detritus of a lost culture – flint from Bushman arrows, used Bushman's candle, quite possibly last touched by a human hand six hundred years before.

'Why are you doing this?'

This time, the familiar question is not voiced by the stern land-God inside me, but Alice, my fellow endurance junkie.

'I –' I begin, then shake the question out of my head.

'That's all right,' she says, in her motherly, empathetic manner that makes me immediately wish I'd had that spirit in my own family. 'We don't always know the reasons why we do the things we do.'

My pact with myself is that I do not talk about this until our two-hundred-kilometre walk to the Skeleton Coast is over. Then I can be sure I have gone the distance. Another reason why I can't seem to speak very well is that I am haunted by something that happened five days ago, or was it six? The days have fused in their eventlessness, in the dire rigour of this landscape.

That night I sleep as usual on a bedroll on the ground. It is dangerous to sleep far from the fire and the vehicles but I don't care; my desire to be alone trumps fear.

I dream we are sand-surfing on windsurf boards mounted with spinnakers; imprinted on them are the names of towns in Damaraland, the Kaokoveld: *Koppermyn* and *Mon Désir, Torra Bay* and *Sorris Sorris*. Such reckless yearning names for drunken hamlets with an Engen station and a bottle store.

There is some formula driving what I am doing on this walk, which I already know is an ill-advised adventure, what I have been doing all my life: an attempt to solve that persistent equation between lavishness and desolation.

Day 5

'What do you think of my Bushman's feet?'

Jan's question goes unanswered. Possibly it is rhetorical, a way to draw attention to the only thing that binds a disparate and increasingly conflicted group of people – our wounded feet. We sit around the fire bundled in our down-filled jackets, but our feet are bare in sandals, our costume of these see-saw days when it is thirty-eight degrees in the afternoon and eight at night.

All night I hear jackal and hyena. Hyena make scoping sounds, almost a coo, like whales communicating underwater. Although it is not smart, given our vulnerability when asleep, I wear a British Airways eye mask against the Cyclops glare of the moon. I wake and shift it to see Orion's boxy eye criss-crossed by shooting stars. The Bushmen said 'we are the Dreamer's dream'; they believed we are being dreamt into existence, which we mistake for reality, by a far more advanced consciousness. It's not hard to imagine a remote intelligence behind these dark skies curdled with constellations.

For the whole day we'd seen only ostrich. By day we walk through eerily vacant Gondwanaland plains. 'Ten years ago this place was teeming, man,' Jan tells us as we rest under the only tree for fifty kilometres in all directions. In the 1800s, homesteaders in the area would be unable to open their doors, walled in by springbok herds that stretched from one horizon to the other. Now weekend hunters come from Windhoek or Swakopmund in portly four-by-fours to finish the animal-o-cide the European settlers started. So far we've scared the living daylights out of a lone mother giraffe with her calf and scattered a few nervous springbok. All animals react the same way to our presence, standing stock-still, staring with blank determination, trying to decide whether we are friend or foe.

Day 4

We walk for seven hours with hardly any rest. The rains have been so plentiful this year we have been walking through undulating sable curtains of grass, so beautiful that I forget to look down. This momentary inattention is what led me to my worrying experience, two days behind me now.

At the top of Messum Crater Jacqueline plays Sinatra's 'Fly me to the Moon' on her phone. A khaki plain stretches to all horizons, punctuated by anvil-shaped drumlins. I take Stephen's poems from my backpack and read them while the wind dries my sweat. I hadn't noticed before how *Return of the Moon* is suffused with dreams – the dreams of the Bushman, of stars and the moon and its steel light, of rain and wind and oryx and eland, of the first green ribbons of dawn that appear on the horizon, far before the sun.

Things that happen today: Alice gets walloped on the head by a metal tent peg which the wind rips out of the ground, tearing the fly tarp, our only protection against the sun in this treeless place, over our heads where it whips, a giant kite. Alice's head pours blood and we marshal one of our precious ice packs, long and thin like those ice cores glaciologists coax from the Antarctic ice sheet, to stem its flow. We are four hours away by road from hospital.

And later, once Alice has been attended to and we are sure she is not going to die of a brain injury/bleed to death, we all sit around the fire, our false camaraderie circling us like a hyena emerging from the shadows. Jan looks at all of us, scoping the group of women he is reluctantly in charge of. His eyes land on me. The expression I see in them is familiar: curiosity, distaste, and an element I can't identify.

Once again, I seem to have drawn a man's ire. It is so easy for me to do this, I do not need to say or act in any particular way, express an opinion that annoys, a political proclivity or a distasteful personal habit. It is something about my energy – perhaps my certainty, or the cumbersome knowledge I have accumulated, through being a writer. Jan's look might be saying: you're a real know-it-all, aren't you? To which I would answer: Yes I am. I have no fear of men's sanction. In fact I enjoy eliciting their distaste; it galvanises an old code within me, one that unlocks the door to a place I can't remember, a zone of not being a woman, of being something, someone, else.

At nine o'clock we scatter to our separate bedrolls. It would be safer for us to all sleep together, wedged between the Landcruiser and the fire, but we do not trust each other enough for that. I sleep with my head against a sheaf of Gondwanaland karst and fall asleep staring at the night sky, the small Magellanic Cloud dissolving in my eyes.

Day 3

Helen is the oldest of us on the trek. She is finding the terrain hard going but is not making it easy for herself. She rejects all our sorties of friendship or offers of help. Each night around the campfire on her iPhone she reads out the blog that someone on the trek kept of this same journey two years ago. 'Dave brought me up to walk behind him,' Helen says, and her eyes mist over. 'I was number two all the way.' The present is no match for the past. I can't help but feel sorry for her. We have perhaps committed the same error, summed up in that old adage: *never go back to somewhere you have been happy.*

I regret coming back here. I want to rub myself out into the sandpaper land. Like scraping myself across its scour, this tired Gondwanaland. A place of pluvials and interpluvials, so many the land itself cannot remember, proto-deserts covered in secretive forests, the sultry Caprivi strip cloaked in clubmoss trees, savannahs punctured with awl-shaped mesas, blood deserts where lone oryx stand in the parlours of abandoned farmsteads.

This land is a violin: taut, resinous, proud: the husband who refuses to divorce, the stepmother who declines to die. The land is hard-hearted, it gives you nothing back but steals your lungs when you are not looking, realigns your allegiances. No place will ever be empty enough, after you have lived here. After Namibia, any other place will be milky and regretful, a foppish dandy. We might need such spaces, where we can be bludgeoned by prehistory, where we can be put in our place. In walking across an original desert like the Namib, we enter into a different state of consciousness. It is impossible to describe, but both the world and the mind gain in dimension at the same time, with the same pace. An old synchronicity.

As we walk, I battle strange lashing panics I belatedly recognise as grief. Stephen's poems unfurl themselves in my head, endlessly combining and serpentining.

Stephen should have done this walk. He would have handled the interpersonal dynamics better than me. He probably would have written an award-winning poetry collection out of it. He knew how to gulp the ground. He wrote that the body of the writer absorbed the message the landscape was broadcasting. Hemingway, about whom he wrote with great insight and affinity, was the embodiment of this idea. He would never have written the books he wrote had he not grown up in the woods and lakes of Michigan. He had absorbed the savagery and the wild intelligence of that place. I would like to be like that. I too loved the wild place I grew up in even if for the first twelve years of my life the wilderness tried to kill me.

Grief is waiting, I realise. Waiting for what? For the moment to return, the moment in which you felt alive, if not loved. Waiting for the moment in which the natural obscurity of life, the dark and the shadows, is illuminated. Waiting for a figure to appear on your horizon.

To come bounding back from the edge, that realm where land and light is swallowed by the mountains.

Day 2

That night, out of earshot in the edge of darkness, Alice pulls me toward her. Her eyes shine moon-bright in the darkness. It is so total her edges dissolve into it. Her hand falls on my shoulder. For a small woman her grip is strong. 'This place is a lion, it is a snake.' She releases me and walks away, where she is instantly swallowed by the night.

I imagine a dual creature: the sandy bread-loaf torso of the lion, the fatal coil of the snake its tail. Yes, I think, this place has the phosphorous of matches yearning to be jolted into fire, with its red deserts that ripple across it like blood spinning from wounds, its flaxen fields supple and golden as the flank of an eland. But its spirit of apparent sobriety masks a capacity for caprice. If places can be said to have a soul, then Alice has nailed it. As I will come to know in the years ahead, the lion is a baleful machine but predictable, at heart, whereas you never know what the snake will do.

Day 1

The spring sun has been up for twenty minutes. We walk across the northwest face of the dune. That is why I don't fully believe in the snake: he should be on the east slope, facing the sun.

I think: just a horned adder. No, it's not—

By then it is too late and my feet are above its plug-shaped head.

I know very well what you are supposed to do when you encounter a puff adder: stop dead, back away. Instead I step over it thinking, oh, that's a puff adder.

When I am a few steps beyond its coil, I stop and tell the others. We track back to the red dune. For a while, we all stare at it. The adder, semi-torpid, seems not to notice or care it has a gallery of spectators.

Jan prods it from afar, eventually lifting it off the ground on the end of Jacqueline's walking pole. The snake unhinges itself and dangles to reveal its length – a full metre, I guess, the biggest puff adder I have so far ever seen. His thick belly is a vanilla bronze. It glitters in the early morning sun.

After he has put the snake down Jan comes toward me; we exchange an uncertain look. He makes to put a single finger on my shoulder, as if to say, *you're bloody lucky*, or, *I'm glad you're not dying right now.* But at the last moment he draws his hand away.

The desert throbs. Green and yellow fever trees, the ceramic rock lit in a red afterglow of a desert sunset. From them emanates a pulsating, swimming heat. At sundown the world changes composition. Immediately, as if it had been waiting around the corner, a cool wind blows. The tiny mopane bees which have by day been feasting on our sweat make for the refuge of the tree whose name they bear. At night, the rocks are sable against the horizon. We go to sleep with the rippling coo of the rock doves in our ears. Above our heads, more galaxies curdle.

'Desire is a great builder of inner space within human lives, hollowing us out, making resonant places we originally thought vacant', Stephen wrote in one of the essays collected in *The Music in the Ice*, which would be his last book. Suddenly inside us there are grand vistas flooded with sun. We say, I never knew this place existed.

I sit up and look into the yellow wind. There is a space inside me which I did not know was there. This space has been prised open by the desert, which demands I fill it with emptiness.

Stephen's voice – as a poet, as a person – still rings in my mind; how its intensity, belied by his languid Capetonian vowels, the humour that always loitered at the edges of even the most sombre sentence, his near-cackle of a laugh. Only when you stop remembering what someone's voice sounds like are they truly dead.

Day 0

The night before our walk begins we sleep on a grassy plain under a full moon. Only yesterday I'd flown from Cape Town in a sandstorm so fierce it threatened to divert our flight. *Don't worry*, the pilot had said on the intercom as we keeled in over the cold ocean. *The engines can take it.*

I wake in the middle of the night from a dream to the gurgle of jackal hunting in the shadow of the Brandberg. In the dream I am in my flat in Cape Town and there is a stranger, a man, in my shower, naked apart from a pair of chocolate-coloured ugg boots. The man is flimsy, urban: a writer type.

I say to the dream in a bleak panic: *Take me back, take me back please!* I wake up in the Namib and the relief is like waking up from a nightmare in which someone has killed you to find you are not dead after all.

There is always a dream the night before the story starts, but we don't always remember it.

PART V

THE BLUE DESERT

August 2ⁿᵈ, 2019. Watamu, Kenya. 28°C

The approach is on one of the Dash 8s from Nairobi that do the run three times a day, in a golden late afternoon. Fifty minutes from the capital, the edge of the ocean etches itself on the horizon. The white rake of waves as they encounter the reef appears as a lace hem to an endless blue garment. The plane arcs over the Sabaki river where it meets the Indian Ocean, creating a vast mud delta where hippopotamus sea-surf and thin fishermen dig for *vongole* in the sand.

Underneath the plane, coral islets scallop the shore. The sea threads in and out of the coast, in eddies, inlets, deep saltmarsh creeks. A peninsula shaped like the lobe of an ear comes into view. The beach is lined with the bristle of casuarinas, those svelte sand-loving pines that grow easily along the shoreline. From above, these look like the sparse spiky hairs on an elephant's tail. Out to sea, humpback and sei whales thread the waves. Sometimes they are visible from the plane, just underneath the surface, grey, submarine-sized lozenges.

Coasts often look exposed, fatally so. But this one is secretive, coy. It beguiles, with the auricle of the creek, its shallow bays and milky fringes and black mane of dead coral. Flatland interior beaches appear, sand dunes curving far inland. These are tidal mangroves, their saltmarsh coated in bougainvillea and coral creeper. All along the stretch of beach outcroppings reach out into jade waters topped by islets – *tombolos*, a rare formation, a kind of pre-island.

The plane banks, arcing toward the runway at Malindi. The surfing hippos and shy battalions of coconut palms recede, into the darkening evening.

*

'It looks like a disaster zone, don't you think?'

I follow his tall, thin figure as we squelch through vats of seaweed spread out on the beach. Sandal straps and brown glass medicine bottles (how Red Bull is sold in India, apparently); an amputated Barbie leg, woven through with kelp; single flip-flops, never to be reunited with their pairs; plastic milk bags; frayed toothbrushes; sheets of bubble wrap; yellow bottles of bleach; pieces of ship's rope in popsicle blue and mint green festoon the coastal bush like Christmas ribbon; a My Little Pony, faded to pale sherbet; a purple and gold shoe like an offcut from a Dolce & Gabbana display; children's hair baubles, all faded and splintered. It is as if the toy section of a department store and a chemist have been fused then detonated on the beach.

The 'he' belongs to my partner, a visual artist who works almost exclusively with what is collectively called marine debris. I don't want to use his name because that would make him both more and less real. And he is not a part of this story, really. Like all people in this book he lives in the shadowlands of the backdrop and not the foreground, where people in books usually belong.

My partner turns all this discarded flotsam into objects of beauty. Often only on close inspection do you realise that the border of a painting is made of toothbrushes, that the mosaic-fringed mirror is cigarette lighters. His house is papered in his creations, so much so it becomes difficult to distinguish them from each other: a three-dimensional geometric wall hanging of decapitated toothbrushes in sun-faded oranges, greens and blues, carefully arranged head to foot, covers one wall. In the corner, light sculptures made from Japanese glass fishing buoys wink. An abandoned child's bicycle has been turned into a lamp and is affixed, upside-down, to the ceiling. Beer coolers and napkin rings of reconstituted flip-flops line mantelpieces. Each surface is coated with mosaics, friezes, sculptures, children's toys, egg cups, also made from the flotillas of flip-flops that wash up on the shore.

His work is a reminder that this environment is under an insistent attack from an armada of sea-bound garbage, insoluble due to its materials. The sheer amount of plastic chemicals is killing sea turtles by giving them corneal cancer. They wash ashore often, dead, their eyes clouded marbles. On fortnightly regular community beach clean-ups, teams of villagers working up and down the beach for only two hours fill the back of a pickup with jute sacks full of plastic junk. I've seen pictures of Midway Island, of course, of the plastic gyres in the Pacific Ocean bigger than many islands. But to see this ten-kilometre-long beach carpeted with the efflux of our lives… The phrase rings through my head like a shrill school bell: *what have we done?*

The history of the East African coast is written in a grammar of monsoons. The wind has defined it, writing a see-saw story of plenty and lack. The summer monsoon, the Kaskazi, which blows from the north, brings salad days: *bain-de-soleil* Italians, blue-white Brits, silver skies and mojitos. Then this beach is Tahiti-pretty, with thin coconut palms draped over the sand, kitesurfers twirling in the lagoon inside the reef and queues for restaurants.

But now, in August, it is the Kusi monsoon, locally called 'winter', when the wind blows from the south. The Kusi is a beguiling name for what is in reality a season of storms. Stern-looking skies. Rigid winds. And plastic – so much that the ocean seems made of it. The rubbish is usually only present in the Kusi, when the current and gyre combine to ship the efflux ashore.

I first travelled to Kenya's coast in September 2012. I came to recover for a few days between my safari guiding course at Lewa Downs and returning to my job in Norwich. As I walked on the beach at Diani south of Mombasa, taking in its huge tidal width, its iridescent sands of mother-of-pearl and lime, I felt an instant affinity with the place. This ocean looks older, I thought, more experienced than others I had seen. I had an intuition: *this ocean has something to tell me.*

I'd never thought that about an ocean before. I had spent years of my life in the tropics. In other parts of the world I'd term what fascinated me was the ordinary melancholia of tropical beauty, and there was plenty of that, here: razor-leaved palm trees, platinum skies, fireball dawns, the thickness of life of the equatorial zone. But the draw wasn't quite any of this, either. It has taken me years to understand that something about this coast, with its meringue sands, the abashed way the waves arrived onshore as if harbouring a tired secret, has the feel of my mental experience, of being inside and outside oneself and the world simultaneously, of being trapped between language and geography, between the zone I privately call *there* and here.

That first night in Diani I wrote down my impressions of the day in a spiral notebook I'd bought at the local supermarket. An idea for a story ignited – it really did fizz out of nowhere – about a white African family who had lived for decades in the languid splendour of the coast, its palatial coral houses, its mosaic of swimming pools and swathes of fluorescent bougainvillea, and of a visitor, a stranger who comes to stay with them, a person who is their relation but who also has secrets to hide. Stories rarely turn up like unannounced guests, in my experience. When they do, you open the door.

There is no doubt in my mind that the inspiration came from the place itself, and not from any story I had been nurturing, looking for a backdrop to assign it to. The ambiguity I felt circulating around the coast, its contradictions – a cosmopolitan culture left to fester by centuries of neglect, its collision of churches and mosques, the wattle-and-daub houses next to gazillionaire mansions, a disparity in wealth that was obscene and glamorised at once – captured my imagination. I heard a familiar invitation, almost a taunt: *write about me*.

The following summer I returned to the Kenyan coast to research the novel this storyline had evolved into, and which would eventually be published as *The Dhow House*. I wanted the landscape and natural habitat of the coast to be a character in its own right. I hoped to diagnose its mystique. But in order to do that I would need to know it intimately, and that requires time.

I booked myself into the only place I could afford to stay for two months: a Christian conservation organisation called A Rocha Kenya, located on the Watamu peninsula a hundred kilometres north of Mombasa. The organisation ran a field station called Mwamba ('rock' in Kiswahili), three quarters of the way down Watamu's pristine white sand beach. There, scientists and conservationists ran ornithology and terrestrial and marine biology programmes.

Living in a Christian commune (despite not being Christian) was very much like being on a research ship, as it turned out, minus the G&Ts: regular meals prepared for us by the kitchen staff, the daily Situation Reports replaced by prayers and reflection, evening group showings of BBC nature documentaries and eccentric expeditions to map the density of the rare Clarke's weaver in the nearby Arabuko-Sokoke forest. I learnt to count flamingos and to bake cakes for thirty people. A committed agnostic, I was leading the prayer meetings by week two. My stay at A Rocha did what I needed it to: it gave me a crash course in the natural environment of the Kenyan coast, which enabled me to write my novel from the inside, as if I was one of its characters, someone who had washed up accidentally on the turquoise shores of east Africa.

Six years later it is September and the Kusi is waning. The skies are hesitant and frigatebirds zoom back and forth along the beach, their stomachs touching the tips of the breakers. It is hard to believe that in two days' time I will sit in front of a fresh crop of students for the University of East Anglia's Master's in Prose Fiction. I live between two worlds now, so distinct it is hard to believe they share the same reality.

My partner's house looks out on five acres of beachside bush, home to monitor lizards, white-tailed mongoose, Sykes' monkeys and bushbabies. At dusk we perch on the veranda's edge. He surveys his domain, a wistful note in his eyes, as if it is him and not me who will soon be parted from this place.

'I feel safe here,' he says.

'Safe from what?'

'Europe, history, trouble. Nothing can touch me.'

'What about al-Shabaab?'

'Oh,' he waves his hand. 'That's a fluke.'

'It really isn't,' I say. 'That's called history. You can't avoid it by hiding out here. Somalia is a hundred kilometres from here and it's not going away.'

'Will you come back. At Christmas.' There is no question mark in his voice.

'I don't know. I don't have any place here.'

He frowns. 'You can stay with me.'

'I meant, I'll never belong here.'

To his credit he does not try to refute this. This is the only place on the planet he can call home, but he does not belong, either. The reason is not the colour of his skin, or not only, but the winds of history, which have shifted definitively in favour of the future, whereas his authority and worldview are of the past. I wonder if everyone of his generation is hiding out, living in this palatial furtive hinterland, avoiding history. When history catches up with the coast, there will likely be no victory or reckoning, no evictions of languorous Europeans drifting from room to room with sundowners in their hands, only the sound of the waves raking the coast, and the distant roar of the ocean, approaching.

Ocean must be one of the most beautiful words in English, with its flowing cascade, the hiss of the 'c', the minor-keyed declension of the *ea* diphthong, one of my favourite sounds in the language. The word sounds exactly like what it represents – fluid, dynamic, vast.

It began its life as a word in Greek: *Okeanos*, meaning 'a very great stream encircling the earth's disc'. Bypassing Latin entirely, it emerged next in Old French, *ocean*, then discarded one of its 'c's somewhere along the way.

The Greeks considered the Okeanos surrounded a single land-mass, with Europe at its epicentre. This imagined Gondwanaland had, of course, long ago dispersed. But the fact that they were right about the nature of oceans in principle, despite their lack of radar, geology, cartography, satellite imagery or GPS, never mind that their navigators hadn't made it past the Mediterranean, is uncanny. A kind of premoni-tion, even, given that this is exactly what defines an ocean, as opposed to a sea: it enlaces continents, flowing contiguously.

'We call it planet Earth, but that's a misnomer; actually its name should be "Ocean",' my colleague at the University of East Anglia, environmental scientist Jeff Price, said to me one day. All the world's continents can fit into the modern-day Pacific. The Indian Ocean is a mid-size body of water in comparison to the Pacific, yet it still accounts for one-fifth of all the ocean coverage on the planet. It is divided into two entities: the eastern and western Indian Oceans. A child of the continental divorce of Gondwanaland 180 million years ago, it was a late blossomer; it has had its present-day shape for only thirty-six million years. But in human terms it is the oldest of them all. People have been traversing the Indian Ocean for at least five thousand years; in the Pacific it is only two thousand, while the Atlantic has a mere seven-hundred-year-old history of human traffic.

There is a rule about oceans: to know one you must see another. I was born on the rim of the Atlantic, in eastern Canada, then I spent years zipping across it in the troposphere. Not until our 2009 Greenland transect when we sailed from Falmouth to Ilulissat did I see it for real. Then, in 2010, I returned, this time in the form of a journey whose dimensions I did not quite compute, except I knew it would be long. We would travel on the *James Clark Ross* from Stanley in the Falklands to

Vigo in northern Spain, sailing up the spine of the Atlantic, a journey few people apart from cargo crew undertake these days.

It is time for me to leave the Indian Ocean, again. Now that my departure is imminent, details of Watamu begin to etch themselves upon me, as if to consecrate themselves: the bulbul who burbles at 5.50am precisely; the chariot sunrises, the sun charging above the near-Equator, triumphant and vain; the wind that ruffles the palm trees by 11am, the sound like sheaves of paper rustling.

In the late afternoon my partner deposits me at the airport in Malindi. Soon I am back in the Dash 8. Evening is gathering. African sacred ibis march through the airfield grass with their monochrome plumage, their heads bobbing up and down as they search for food, like miniature Texas oil drill heads.

I gather my body, tensing for take-off, as if I could keep my limbs together by sheer will, should something go wrong. The plane levers itself into the sky and the pilot banks hard to the right. We follow the golden serpent of the Galana River as the sun sets. Whenever I leave a place it begins to fragment and pixelate before my eyes. Islets of green. Horseshoe-shaped riverbends with gigantic sandbars laced with crocodiles. Plains of arid solitude. The barnacle of Kilimanjaro, no snow.

For once I do not open my computer to work, but keep my faced pressed against the window as it turns cool with the stratosphere, looking at the tiny homesteads in the bush beneath, squadrons of trees, swimming pools, farms, thinking how it looks like Eden, that charged word I have sworn to avoid, when writing about the landscapes that move me, how it probably is.

Two hundred nautical miles ENE of Porto Alegre, 28.41°S, 44.74°W

Departures

April 30th, 2010. Falkland Islands. 6°C

At the top of the *JCR's* gangway stands Richard the Purser, bald head gleaming in a shard of Falklands sun. I shake hands with Captain Jerry and Duncan the chief engineer, who I had last seen seven months before in Greenland. We trade the standard ship-convict jokes. *Back for more punishment?*

It is the southern hemisphere autumn, the tail end of the Antarctic research season. Able Seamen rush past us on the deck, their faces creased with fatigue. By this point the entire crew have been working for three months day in, day out, in the Antarctic. We climb the narrow jungle gym stairs to the accommodation decks. This time I am assigned cabin 10, just a few cabins down from my last home, cabin 4, on our Greenland trip. The ghost of that journey and the bleak crisis I suffered in Ilulissat tugs at me every time I walk past its door. Not once on this trip up the Atlantic will I enter it.

We all gather in the bar for the pre-departure briefing. If anyone is surprised to see me aboard for the voyage they do not let on. The cooks, stewards, ABs all treat me like a piece of the furniture, issuing gruff *nice to see you again*s.

Stanley Sound is visible through the bar windows, baleful and moody. The moments before departure by ship on a long journey are different from any other leave-taking. The excitement and finality of leaving is tinged with a resolve; we will be alone for a long time, away from the ordinary world. On a ship, anything can happen. Your life is not your own. You cede it to the ocean – and the captain, of course. The sense of shared resoluteness creates an intensity that has nothing to do with our connection to each other as people, but from a joint vulnerability.

We will leave Stanley that evening at 6pm with the tide, Captain Jerry announces. It is not ideal to depart in the hours of darkness, he adds, but the sea state is relatively calm, and he is keen to get the ship underway. 'It will be a long trip home,' he warns. 'You'll need to keep occupied. Duncan will be handing out a list of tasks tomorrow morning. We'll be cleaning the ship from top to bottom.' Everyone groans. This means struggling up and down stairs with buckets of bleach, scrubbing pallets in the burning tropical sun on the aft deck, scraping rust off the ship's scarred hull, painting the conning tower. 'You'll be grateful for the work,' Jerry tells us. 'You'll see.'

Before plying the ship away from the wharf we are called to do an abandon ship drill. These are easier to do when the ship is static, as it involves lowering the lifeboats. There is just enough light. Out on deck, we watch the sun approach the horizon in a greyscale sky. A solemn escort of cormorants and gulls foams around us, as if they intuit we are about to set sail. I take a last look at the Falklands. I don't know it yet, but this will be my final trip to the Islands, possibly forever. I drink in their sombre skies, the matted dark green of the Diddle-dee, a local grass, the stout forms of upland geese bedding down in it for the night.

After our drill we all retreat to our cabins. On a research science cruise, we would end up in the bar, talking for hours about our mission and its importance, circuiting between games of Trivial Pursuit and backgammon. But I can already sense that this trip will be different, in ways I had not quite taken on board.

I have a personal mission on this trip, which is to write about what a decade from now will come to be known as slow travel. By the time we reach Vigo in northern Spain we will have hauled ourselves

11,470 kilometres at bicycling speed. I had never done such a long journey by ship. I knew the journey would be monotonous. 'You won't see anything,' people I knew from my time in Antarctica and who had done it warned. I took this to mean cities, harbours, destinations. Another Antarctic old hand had told me, 'There's nothing to do.' But wasn't that what I sought? For the first time I wanted to be forced to be as still as one could be while in motion, to do nothing other than to feel the tumbling orbit of the planet, so that I would know the true girth of the world.

The Ninth Wave

As we lose the cover of the Falklands archipelago the sea state stiffens. I go up to the bridge. From there we can see black-browed albatross, curling in and out of the waves. But they will soon disappear, the ABs tell me, as if a button has been pushed. Albatross rarely venture further north than the 35th parallel. Their wings are unsuited to flapping, and the calmer air of the subtropics requires lots of that.

The Southern Ocean is the wildest, most fierce and least-known body of water on the planet. The sea is hardly ever blue but obsidian, that black-grey transparent volcanic glass with a vein of amber. Oceans have characters. There's a mercenary note in this one. It will fight you for no other reason than because it can.

The following night as we approach the coast off Rio Grande do Sul we encounter a storm. We had seen it coming on the radar, which revealed its broad outline but unlike the more sophisticated radar on airliners it gave no clue of the storm's intensity. We passed through a curtain of torrential rain. The sea state changed instantly. The waves lost their rolling, iterative pattern. Now they were swelling, jostling, unresolvable, an infinite family argument.

From the bridge we had a Panavision view: white water leaping out of darkness as lightning cut the horizon, strobe-lit, the sea momentarily turned the white of Antarctic ice sheets, pallid, solidified by a transfer of heat and power. The peals of thunder were sharp, telluric, like metal being torn apart: *scree, scree, scree*.

I decided to try to sleep. I lay in my bunk, feeling the motion of the ship through my back. We were pitching hard and occasionally rolling. On a ship pitching is the standard movement. But sometimes, due to prevailing currents and winds, the ship becomes subject to sideways forces. When a wave catches it 'on the beam' as the mariners say, the world becomes suddenly lateral. Every ship has a degree to which it can be pushed over before capsizing. The *JCR's* maximum roll is displayed on the instrument panel on the bridge: thirty-eight degrees. This is calculated by physicists and engineers. The officers tell me no-one really knows how much roll the ship can take until you're in the situation.

This Southern Ocean also generates the most so-called freak waves on the planet. More is known about these now, because they can sometimes be spotted and measured by satellite. Ships the size of the *JCR* are designed to cope with waves up to a height of fifty feet (fifteen metres). I remembered talking to Captain Jerry about this on our Greenland voyage as we sawed through the suet of the north Atlantic. 'But after fifty feet – well, you're on your own,' Jerry had said. 'Two large ships sink every week on average, somewhere in the world. They just disappear,' he told me. 'Nobody knows what happened.'

'What do you think sinks them?' I'd asked.

'Rogue waves, methane bubbles, structural failure.'

'What happens when a freak wave strikes?'

'It depends on the ship and the wave and the dynamics between them. The ship may take on so much water that it sinks, so that's pretty straightforward. Or you could capsize. If it strikes the ship end-on, one end will dip down into the trough of the wave then be hiked up. This has enough force to break the ship's back.'

I replayed the conversation with Jerry to try to get to sleep – probably not the best tactic. But at some point I must have drifted off. When I awoke I found I was standing upright in my bunk. I fought my way out of the sheets and tried to stand on what had been the floor but was now the wall. My hands shook slightly as I hauled on my trousers. I opened the door to my cabin and was immediately flung to the far side of the passage. On the stairs I had to grip both the banister and the wall to keep upright. I had been in bad weather before but this was different. The ship felt rebellious under my feet, out of its depth.

When my eyes adjusted to the darkness on the bridge, I made out the silhouette of Henry, the first mate, at the helm. He was standing beside the driver's chair. Beyond him, the giant bridge windows appeared to me, not for the first time, as a film screen. The totality of the bridge windows gives a feeling of command. Surely nothing bad can happen, you think, because we have such a good view. But they also turn the ocean into cinema. It doesn't feel entirely real.

Henry was doing this trip as an agency worker. In his normal job he worked oil supply ships off the coast of west Africa. He was affable and approachable, what was referred to as a straight man in Antarctic circles. This had nothing to do with sexual orientation, but with moral approach; a straightforward person, without 'side'.

I went to stand on the other side of the driver's chair. We watched as the *JCR* mounted a crest of a wave, winching herself up its slab-like bulk. The image came to us a split second before the physical force. The ground beneath our feet rose, exerting pressure on our solar plexus. I had always liked this part of the pitch on ships, the floor coming up to meet you, the G-force compacting the body, squeezing your joints, like the thrill of a rollercoaster ride.

The ship finished its ascent of the wave. The sky emerged, beyond the wave but part of it, as if constructed out of ocean. At the top of the crest the ship hesitated, a dancer uncertain where to place her balance and torque. Then she proceeded down the slope of the wave, as calmly as a car on a hill. Another pleasing feeling: the declension of a ship falling into the curve of a wave, that pulpy stomach-slide into gravity, not unlike the feeling of falling in love.

Suddenly a force slammed us from the side. I clapped onto the nearest instrument console to keep myself upright. I looked at Henry for reassurance. He was staring at the instruments. The capable hum of navigation equipment and the VHF radio were at least as reassuring as the officers who had spent years in naval college learning to not sink ships.

The ship swung back upright. The tea kettle fell off its shelf. I ran over to put it right. 'Don't bother. It'll just keep falling over.' There was a snap in Henry's voice. I returned to stand by the driver's chair. I held onto its left arm; Henry, the ship's controls in front of him, gripped the right. The wind battered the bridge windows. I could see them flexing in their frames.

'How high are the waves?'

'Around twenty-five feet,' Henry said. 'Maybe thirty. Not bad.'

In any other country's lexicon *not bad* would be good. I'd been in Britain long enough to know what it really meant.

We watched the sea roll toward us again. Henry had turned the searchlight on, to better see the swells. Usually at night waves were the colour of graphite. But these waves glittered like metal. Then, something – an urge, an energy – seemed to coalesce itself out of a dimension that was and was not the sea. Our eyes struggled to it make it out. Aha – a wave, emerging from the grainy night. Just another wave.

'Hold –' Henry didn't get a chance to even say *on*. We watched the foremast disappear. The windows became water as the bridge entered the wave. My knees ground painfully. I was shrinking, getting shorter. Then we were inside the muscle of the wave. It was tar, algae-green, striated. I thought: *marble*, that green-pink kind, Italian. What is its name? My marble vocabulary has never been great.

We were both pitched forward, almost off our feet. Together we gripped the navigation console. I was worried I was going to be thrown into the window. My knees buckled as the floor of the world fell away underneath us. We were suspended in the air, figures in action-movie wind tunnels, temporarily horizontal. It was a long two seconds. Maybe three.

When the ship hit the bottom of the wave's curve there was an immense bodily shuddering. The ship began to speak, something it did rarely. It said *humm, ho, hargh, hooo*. Then the foremast emerged, the world revealed itself again, a dark sky with holes punched out of it by stars.

Behind us I heard the door click open. A white shirt made its way toward us out of the darkness. 'Everything okay?' Captain Jerry said. 'Oh, it's you driving the ship, is it? I always say give the ship to the writer if you want a solid night's sleep.'

Henry and I tried to laugh at his joke but the sound got caught in our throats, where it gurgled.

Jerry and Henry stood behind the instrument panel, conferring on whether to change course. The ABs came out of the shadows; they had been scanning the ocean from either wing of the bridge. Usually they were jocular, up for a laugh, but tonight they were silent.

'Was that a ninth wave?' I asked. Every nine or ten waves the sea generates one with more power. People think it's sailor's lore but it's entirely true.

Henry answered. 'No, that was a nine-hundred and ninety-ninth wave.'

I am not sure how long I stayed on the bridge but it must have been hours. We all scoured the sea in case it generated another wave-mountain, but it grew progressively calmer. At five in the morning I bade goodnight to Henry and Jerry and fought my way into the sheets on my bunk – Richard had some sort of Army way of making beds that made you give up after a couple of tries and lie on top.

I lay with my eyes wide open for a long time, imagining what would have happened had the wave tipped us over. If we were lucky we'd be in the lifeboats, vomiting our dinner into their fibreglass alleys, stuffed in survival suits. Or would we have sunk like a stone? It was possible that we were never in any danger; Henry had become tense, but he did not seem to feel real fear. That had a smell on ships, I had learnt: oily, scuffed around the edges. It was unmistakable.

I thought of another of Jerry's comments on our Greenland trip the year previously: 'things tend to go wrong fast on ships'. One minute you were digging into the stilton and a bottle of Beaujolais and the next minute you could be in sub-zero waters struggling to breathe, he said. Ships were a condensed lesson in the fragility of life; they looked so capable, even eternal, but a single wave could destroy your world in seconds. On them I felt safe, I felt doomed, and they were indistinguishable.

Crossing the Line

Our lives sedimented into routine. Soon we were synched to it as if ordered by hormones: breakfast at 7am; morning smoko (tea break) on the bridge at 10.30; circuit training in the science hold at 11am; lunch at 1pm; afternoon smoko at 4pm; dinner at 6.30.

Even if my body obeyed the code, I couldn't get my mind to slow down to ship speed. I tried to train myself not to expect event, nor await occurrence, to accept that nothing that was not us would happen for weeks, unless we were unlucky and were wrecked. That is the reality on a ship: nothing is the best thing that can happen.

Where *are* we? I asked myself this question every day in front of the GPS readout. It showed us as a small cross in a sea of squiggles – depth-contours. In each contour was a number in metres: 243, 1006. At the bottom of the screen was the readout in 'lat/long', as the officers abbreviated it. So I knew where we were laterally and horizontally, that was not the problem. I experimented with ways of thinking about our location: we are *off the coast of Brazil*, we are *near Rio Grande do Sul*, we are *in international waters at the bottom of the Atlantic Ocean* or *above the Argentine Abyssal plain*. I was measuring our existence in relation to land because I'd been doing that for my whole life. But we couldn't be said to be *there* either. We were somewhere, I decided, not relative to anywhere else. We were on our own.

I went out on deck to look at the night sky. Clots of stars. A racing, hungry moon. The sky was darker and denser at sea, the stars felt more present, but also as if they might suddenly combust. Silver light fell into our eyes from them, like liquid metal poured from a beaker in a chemistry experiment. On the bridge the officers spotted bioluminescence. We all rushed to the fo'c'sle to see it, groping in the dark along the gunwale. The effect was as if the Northern Lights had been submerged and dispersed in the ocean: an iridescent wave, like titanium or petrol. A sea of green internal fire.

As we advanced northward the air warmed. The *JCR* revealed herself to be not built for heat. The 'air conditioning' only managed to shuffle warm air around our corridors. In the UIC lab I found a fan abandoned in a corner and plugged it in. It whirred as I typed, my chair placed carefully on the diagonal to reduce as much as possible the rolling motion.

We had patchy internet on board. *The Guardian* took painfully long to emerge on the screen. I'd brought only a few books to read with me and the ship's library, a broom closet lined with shelves, consisted of Wilbur Smith neo-colonial epics and thrillers with black and silver covers inside which women were murdered in service of narrative drive.

Monkey Island, as the outdoor observation deck above the bridge was known, became my television. I stationed myself there twice a day, after lunch and in the evening. In between I walked circuits on the aft deck, rescuing flying fish that stranded themselves there, shocked and gasping, a fluttering carpet of exhausted miniature angels.

Then the tang of sea air was replaced by the taste of ferns. The air softened, the wind relented. The sea seemed to thicken. The zooming porpoises that we'd seen intermittently since departing Stanley left us. The only other sign of life was a conveyor belt of boobies, gannets, boobies, gannets and occasionally a tropicbird – the fighter pilots of the oceanic realm, appearing as a white arrow high in the sky.

On the aft deck, Clem, one of the ABs, sat in a plastic chair, his Antarctic-pale face angled upward like a satellite dish toward the sun. I dragged a chair to join him. We sat in silence awhile, looking at the ocean. The sun was tangerine. Shafts of light penetrated the sea's thickness. The only sound was of us, moving through the ocean. It sounded like a knife flensing something thick – rubber, or skin.

'I remember when these seas were thick with fish – marlin, bill-fish, dorado,' Clem said, eventually. 'Now look. I turned vegetarian a couple of years ago,' he said. 'It's tough going on the ship – Ash doesn't even know the word.'

Ash was the ship's cook who dished out tuna bakes, chicken Kievs and shepherds pies. 'It's impressive that you can hold the line, Clem.'

'Oh, I just couldn't live with myself otherwise.'

'I've had this strange feeling lately – like there are ghosts in the sea,' I said.

Clem's heavy-lidded eyes flew open. 'Ghosts?'

'Well, something. Like an absence.'

'Oh yes, there's enough ghosts,' Clem said. 'The horses they threw overboard, the Spaniards caught in the doldrums when they ran out of water. The slaves they threw in the ocean.' A pensive expression settled on his face. 'That's one reason I went veggie. I just thought: all these dead animals you're eating. I couldn't stand being haunted. Their spirits are inside you.'

Even though I had never seen it as Clem had, the ocean did feel empty. Sailing through Antarctic and Arctic waters had always been like being afloat in a pelagic stew: blue, humpback, fin and pilot whales, marlin, toothfish, dolphin, schools of herring that looked exactly like piles and piles of sesame seeds. But here in the subtropics, the ocean had taken on an unfamiliar silence; there was a withholding note in it, an eerie echo.

As we inched up parallel with the coast of Brazil the skies become cloudless blue tarpaulins. The air was hot and lurid. The ship's air conditioning failed entirely, so we abandoned our dinner table dress code and arrived at lunch wearing singlets and shorts, our brows greased with sweat.

I had come aboard to experience a long trip by ship over the open ocean. It was a deliberate plan to immerse myself in a dimension few people ever get to experience. But it was proving harder that

I'd thought, to see only the sea, these empty blue barrens, the ocean that drinks in the sun that falls again and again, an axe striking in angry insistence.

We near the Tropic of Capricorn. I am outside on deck, scrubbing down pallets from the alcohol store, the smell of stale beer making me nauseous in the morning sun.

The word tropic comes from *tropikos*, in Greek, which has its origins in the verb *trepein*, to turn. At the Equator the planet spins at 1,660km an hour. (At the South and North Poles, by contrast, it does not turn at all.) Never before have I been on the Equator while at sea. I have the impression that instead of hurtling faster, the planet's plunge-orbit has stalled. The sea is thick and sluggish. The skies have the look of skies that were once on their way to somewhere, but have become fatally distracted.

Empty of Antarctic cargo, the aft deck is an enormous oil rig platform. The crew look heroic, tiny figures leaning into the infinite ocean. If something went wrong and one of us fell in, we might have a chance of rescue, but only if there was a witness. An image implants itself in my mind: I am in the ocean, treading water, watching the ship recede into the horizon. The back of my throat is already parched, the tropical sun sears my head. Overhead, gannets regard me coolly. The image scares me because part of me wants to do it: jump ship. Just to see what would happen. I've never had such an impulse before. It feels like a kind of vertical vertigo.

After two weeks at sea, Brazil is still there. I begin to wonder if we had entered some *Rime of the Ancient Mariner* zone where we are doomed to sail the coast of Brazil forever. Finally, just north of Recife, we curve away. We will cross the ocean by hopscotching along a chain of volcanic islands: Fernando de Noronha, an amoeba-shaped island off the coast of northeast Brazil, to St Peter and St Paul, mere specs in the Atlantic, then Fogo in the Cape Verde chain.

Suddenly the air smells green. Within an hour of changing course away from the coast of Brazil, brown boobies fly in and perch on the anemometer. They regard us officiously, like harbourmasters come to pilot us. One remains all the way to Fernando de Noronha, as if catching a taxi.

The clouds assume the anvil shape of the tropics: svelte at the waist, rising to flattened, lunkheaded termini that brush the troposphere. At 5.15pm precisely the officers of the watch and whoever else was up for the sunset channel assemble in front of the bridge windows. 'It's always like this in the tropics,' Henry says. 'The joke is, "who needs DVDs when you've got Sky?"'

There is no doubt we really are on a hurtling orb. Night after night we watch the sun charge into the curve of the planet, fast, as if the sun and the land are squaring off against each other in battle for the end of the world. The clouds, so recently abandoned by the sun, flare from within, lit by fuchsia. A roseate aura floats through their girders. Just before the sun sinks into the horizon the clouds acquire a lemon nucleus near their summits, as if striving to retain a power source.

The sun disappears, dissolving into the lip of the land. Its vanishing ignites an amber firestorm. This explosion imprints shapes in the sky: a feather, a hammer, a conch shell, the silvered silhouette of a fish. These shapes are stable only for a few minutes before mutating. The clouds press down on the sea, narrowing the depth of field between them until the horizon begins to pulse, pounded by an invisible instability. White-gold flutes of light fan into an indigo sky, knife-straight.

But soon these throb and dissolve, fire embers recalled to their source, now plunging westward across the Amazon and into Peru. A red meridian is etched across the sea, like a brand, cauterising the night. The officers call all this the 'aftershock'. And it does this job, of aftermath, hurrying us into a pensive, chastised mood at the sheer lavishness of its desolation.

Night after night in the Intercontinental Tropical Convergence Zone we line up on the bridge to watch this spectacle like members of an animist cult. Every evening it is the same but different: different colours, combinations, different cloud architecture. But each sunset retains the shadow of the one the night before, like a ghost returned to haunt itself.

At 4.50am we cross the Equator. *Did you not feel the bump?* is the joke on board. The bump is yet to come, in the form of the ritual Grime and Punishment ceremony. Along with the outgoing Antarctic base doctor and an assistant electrician, I am the only person on board who has never crossed the Equator by ship before.

The three of us are rounded up and dressed in Droog-like overalls. We are marched to meet King Neptune, played by Sly, an AB from Liverpool. Sly interprets Neptune as a drag Pippi Longstocking. Two orange braids fall from his cap and from somewhere (how, we wonder, on a ship?) he'd procured purple striped tights. It is Sly's costume, more than the fermented cabbage we have dumped over our heads on deck, that makes us throw up. Once the vomit has been hosed off we take three showers in a row, but the fume of rotting vegetables lingers for days.

After the ceremony the energy on board crashes. People gather in listless clumps in the bar. There is nothing to look forward to now, apart from arrival, and that is still two weeks away. Every conversation we might have held has already happened. Everyone is crossing off the days. Anywhere else, the result would be ordinary boredom. But on a ship in the middle of the ocean the lack of common purpose creates an impasse that is easily turned in any direction.

A suspicion of me emerges in this vacuum. No-one says anything but it reveals itself in the Tudor Court dynamics of a ship. Who talks to you in the bar. Who sits next to you at dinner. The people who knock on your cabin door. The people who meet your gaze.

I am the only one taking this trip voluntarily. Everyone else is being paid to endure it. They are bored, and bored people's gaze is magnetically drawn to the weakest link in the group. Ever since I had walked up the gangway they had been wondering why I would give up four weeks of my life for what is to them a long and dull train journey. They'd seen me writing, of course, tapping away on my laptop in the UIC lab with the chair bolted to the floor, but this was not enough to convince them of something they wanted of me which I did not understand and so could not deliver.

We no longer congregate on the aft deck to watch the stars at night. Or if we do, we pitch our plastic chairs far enough away from each other to not have to talk. In the evenings everyone retreats to their cabins to watch the standard shipwreck documentaries on DVDs. I take to spending long periods on the bridge, the only place on the ship I feel tolerated. The truth is, I can hardly write, other than to keep a sluggish diary of the trip. The foreswearing-of-time atmosphere thickens the air in my head. Far from a romantic quest to cross the seas, I realise I am engaged in that most squalid of activities, the using up of time.

Dark Ocean

We are 330 nautical miles east of Belém, smack in the middle of the Atlantic. I hang over the gunwale on the aft deck, my favourite position when I don't know what else to do. George, an AB who has been with BAS for thirty years, joins me. Ship spray coats our arms. The water is skin temperature; we don't even realise we are wet.

We stare at the sea surface. It is thick and iridescent, like taffeta.

'What are you looking for?' George asks.

'Anything. I can't even see jellyfish.'

'We used to see all manner round here on passages north,' he says. 'You name it: bluefin tuna, dolphin, marlin, mahi-mahi. Now, well you can see for yourself.'

'Yes, Clem was telling me. I'm trying to imagine what it was like.'

'Think of a pot of boiling water,' George said, scanning the ocean with empty eyes. 'But the bubbles are fish and dolphin. The sea was *stacked* with them.'

Mentally I sketch in my impression of what George has seen: wheeling waterbirds foaming, stiff ruffs of sailfish, matted shoals of tuna, their blue-sunk bodies, the gold-indigo shimmer of dorado, gangs of gannets.

'I wish I'd taken pictures,' George says. 'I thought it was normal, that it would always be there. But now it's a ghost ocean.'

Ghost ocean. The phrase hovers in the air between us. It has the same insistence as the waves we plough through. This sea is a Claude glass – one of those dark, convex mirrors the Victorians carried for gazing sublimely at English landscapes. It seems reluctant to let us through. Is it possible the sea could be thickening? Captain Jerry had told us about methane bubbles. These sometimes escape from the

seabed, rising from the mysterious Abyssal plains kilometres underneath us. Methane bubbles are suspected in many of the cases of disappearing ships, Jerry had told us. Heavier than air, methane destroys a
ship's buoyancy. 'They go down like a stone,' Jerry said. 'So fast you
don't find even a single lifejacket.'

Right on cue, I start to have anxiety dreams in which I am on the
ship, we are moving through the ocean, and the sky disappears. Suddenly
we see a black wall. The ship pitches forward. I am thrown on my knees as
we nosedive through a vacuum. We are heading down, into outer space.

Two hundred nautical miles west of Cabo Verde we crossed an invisible meridian. The hot, murky sea we had sawn through for two weeks
brightened into the sharp azure of cornflowers. Henry confirmed it.
'We've picked up the Equatorial Counter Current, and it's going to the
same place we are for a change.'

To the east, Africa advanced after all; for days it had seemed as if
the continent were moving inexorably away from us. The GPS showed
us parallel with the Gambia and Sierra Leone. *Africa*: the continent
I left only six weeks before on the Malaysian Airlines flight from Cape
Town. It felt much longer, like I had switched planets.

For half a year my life had been about distances and emptiness:
open skies, seas, vacated horizons: the austere plains of Namibia,
the maw of False Bay south of Cape Town, the jerry-built desolation
of the Falklands. Now I was about to return to Europe, to cities,
newspapers, espresso, friends, mobile phones that actually worked.
I would no longer have the protection and purpose of motion and
writing to power my life. I would have to sit still, and stillness would
bring with it the sullen-voiced overlord in my head. For most of the
ocean voyage the internal overlord had struggled to find me and
deliver its rebukes. Maybe it didn't have GPS.

But now it is back. *All this motion is a consolation prize. You have no life,
no children in gingham uniforms, no ginger cat, not even the birds look at you as you*

pass, you are husband-less, you have the arrogance of those who don't have to share ice cream or toilet paper. You never had the guts to be fully human.

The voice has really hit its stride now. *Meanwhile you assign yourself missions to give your life purpose and structure, but what will you write about this voyage, which can only be described as boring? Everyone on this ship is just waiting to get home. They don't understand what you're doing here. What is there to write about anyway? Your hours walking circles around Monkey Island counting waves?*

That night, in part to escape its harangue, I drag out a plastic chair onto the aft deck to watch the stars. I mean to be alone, to have time to consider the sudden stab of unhappiness with myself. But as happens so often on a ship, there is no being alone – even if you try to hide in your cabin, anyone can knock with news of a safety drill or more cargo pallets to be scrubbed.

Before long I'm joined by Ash, the chief cook. He has ham hock forearms and bright blue eyes. He looks like what he was in a previous life, a butcher. But, like almost all the people I meet at sea, his appearance is deceiving. He's well read – currently he's tackling *The Count of Monte Cristo*. He tells me he lives in a trailer in Ramsgate. He is getting a divorce, so he doesn't even have enough money for a new fridge, he says. He tells me the trailer park is a great place to live, that The Krays' niece runs the local pub. 'She's fine, as long as you don't cross her,' he says. He lights a cigarette. The ember follows the ship's gentle camber, undulating in the dark like a single firefly.

'What do you look forward to most about getting home after months on the ship?' I ask.

'Ah, that would be Sir Bobby. He's my spaniel. He was a bit poorly last time I was home. We had to take him to the vet. As he was coming round from the anaesthetic the vet said, "well done Bobby, here you come," and I said, "no, that's *Sir* Bobby to you." The vet said, "ok,

Sir Bobby then," and he came round straight away. He pretty much jumped off the table. That's how much he knows his name.'

I tell Ash how last night I couldn't sleep again in the Principal Scientific Officer's cabin, to where I had moved on the advice of Jerry and Henry a few days before. 'I find it spooky at night. Nowhere else on the ship bothers me.'

'That's interesting,' he says. 'I never get a good feeling from it myself.'

'But how can there be ghosts at sea?'

Ash turns his keen blue eyes on me. 'The sea is full of the dead, especially in these horse latitudes. You do know slaves and horses were chucked overboard, don't you?'

'Yes, I do. But you never know where they are. The ocean is so big. Do you think it's possible that ghosts clamber onto passing ships?'

He gives a gunfire burst of a laugh. 'Tell me, what's not possible in this world?'

'Hmmm,' I concur.

'So what did you do when you couldn't sleep?' he asks.

'Don't tell anyone, but I went and slept on the sofa in the bar and got up at five a.m. before anyone could discover me.'

'Good plan, that. I'd probably do the same. Why don't you just move cabins?'

'No, Henry and Jerry gave it to me, it would seem ungrateful.'

'Well, good luck,' Ash says, hospitably. 'Let me know if you figure out who it is.'

The ship hisses through the ocean. Stars unfurl in sheets of muslin. We crane our necks back in our chairs. We are level now with Mauritania and Western Sahara. The sea seems to be in a hurry to get somewhere. It tosses the ship aside, discarding us with each trough.

'It changes every day out here,' Ash's voice curls out of the dark. 'I spend all my time at sea looking forward to getting home, but when I get there I miss it. So there's no winning.'

'What do you miss, exactly?'

He gives another sharp laugh. 'Freedom.'

Two days later we steamed through Cabo Verde. Fogo appeared as a black shadow on a moonless night, the cone of its strontian volcano piercing the sky. Then up and up, east of the Canary Islands. Further north we would pass around a hundred nautical miles east of Madeira, Jerry told us – 'not close enough to stop for a sherry, unfortunately.'

From the aft deck I saw porpoises, the first cetaceans I'd seen since the seals and Commerson's dolphins who had accompanied us out of port in the Falklands. The sky and sea were a crisp blue, almost the same colour.

Our journey had taken us through zones – Southern Ocean, subtropics, the ITCZ, as mariners call the tropics, Africa, then – what? Islands like the Canaries, where suddenly there is IKEA and mega-ferries bearing lettuce from the Spanish mothership. *Europe*. I turned the word over and over in my mind. I'd never given it much thought. It felt like a concept and not a place.

Later that day Henry announced that we were parallel with Madeira, and sure enough, the air took on that green smell again. As our destination came into view on the GPS Richard stopped putting fresh fruit in the bowl in the bar; two browning bananas sat there, uneaten. As well as fresh fruit we'd run out of sparkling water and our stocks of the mediocre Uruguayan beer the ship took on board months before in Montevideo were perilously low. Within days we would re-encounter the cash economy, and with it, choice. Come to think of it, none of us had any cash. We hoped the ATMs of Vigo were working.

Two days later we made our approach at dawn. The sunrise was a pink fireball. We all went up on deck to witness the city coming upon us, awestruck at its stasis, as if it and not the ship were the floating dirigible. We gazed with startled wonder as land-life – cars, people walking dogs, trees – enlarged itself from toy-town dimensions, closer and closer, until it ate up our horizon.

Suddenly the ship was alongside, the pilot was filling out the paperwork with Jerry on the bridge, our passports were inspected, and we were free to explore the town.

It was a hot day, twenty-nine degrees. I found a small grocer and bought an orange, two non-brown bananas and two yoghurts – things we hadn't had on the ship for the last two weeks of the voyage. The bill came to €5.15. I possessed a piece of runic parchment, which I handed over, out of practice and so a little too slow, to the shopkeeper, who seemed unfazed. Vigo was a port town, and he was used to people coming off ships having forgotten completely about the abstraction known as money.

I walked the streets, allowing a different loneliness to settle inside me. I was thrilled to be able to put one foot in front of the other and for this motion to take me somewhere, instead of my orbits of Monkey Island. But I was unused to being by myself. I could have an accident or be kidnapped and no-one would know! On a ship, someone always notices if you are there. In that sense it is somewhat like a family. If you disappeared, there would be a you-shaped hole in people's shared reality. As I walked around Vigo's quiet morning streets it alarmed me that I lived almost all my life in this state of vulnerability without noticing it.

I found a wedge of shade and stood, trying to compose myself. Wasn't this alone-ness the exact state of being I'd missed for nearly four weeks now? I had been alone in another way, at sea among people I did not really know, some of whom had come to actively dislike me for reasons I could not fully understand, but which were probably procedural (*what is this writer doing on our ship wasting space?*) rather than personal. Yet even if I had no desire to see them again, as I roamed the streets of Vigo their voices rebounded inside my head. I heard the

Second Officer's clipped, mechanical tones most, even though he had been my least favourite person on board.

It was time to return to the ship to collect my bags. I would fly from Vigo to Madrid and then to London that afternoon. I walked back to the dock, up the gangway, passing through the red carpeted corridors lined with framed photographs, all heroic selfies: the *JCR* in pack ice; the *JCR* in the blue waters of Mauritius on a science cruise; the *JCR* snug in the crater of Deception Island off the Antarctic Peninsula. I drew my hand along her rust-pitted railings and leapt, with the same jungle-gym dexterity I observed in the ABs when I first came onboard in the Falklands, up and down the tween-deck staircase.

Two hours later I was walking in the other direction, down the gangway and out onto the dock, and into the rest of my life. I said goodbye to Jerry, to Henry and the other officers, packed my bag in the Principal Scientific Officer's cabin and bade goodbye to the ghost, whoever they were.

The journey had been a pre-forsaken one, taken on false assumptions: that I would witness the teeming plenitude of the open ocean, that I would make friends among like-minded people, that I would write something elegiac and profound about the monumental solitude of the earth's seas. Now, time has congealed the journey, as it often does, into a single image. I am hanging over the gunwale, my arms coated in sea spray, trying to resist the vertigo of the ocean. I can see myself floating on the ocean's black meniscus, waving my arms wildly as the ship advances into the horizon.

The ocean is darker, so much so, than I thought it would be. Planets are buried in it. They are called grief, regret, euphoria, terse interior spheres, charged, primal words to build worlds around. That is what we are doing, always, as writers. It is a utopian project, to construct worlds that ought to have, should have, could have existed. The moral list in fiction is toward the possible. It is an ameliorative impulse.

All those possible planets would be lost, were I to toss myself into the ocean's black glass. Maybe it would not be such a pity. No-one – certainly not me – would ever know. Some people feel the allure of the fall. The sensation that overtakes me on the ship is something similar. If I had thrown myself overboard, could I neutralise the unexpected darkness I had – we all have – fallen into, via our assault on the world-that-is-not-us? It would be a kind of sacrifice. Woyzeck, the main character in Büchner's eponymous play, says, 'Everyone is an abyss. You get dizzy if you look down'. The dark ocean is there, not agitating across the planet, but inside us.

PART VI

BUSH OF GHOSTS

March 11*th*, 2020. Norwich, UK. 14°C

The seminar room looks out onto a large Scots pine, which casts pink-sheared shadows across our desks. At the far end of the classroom a portrait of W. G. Sebald regards us with his cool blue eyes. Sebald is the custodian spirit of our department of English and Creative Writing at the University of East Anglia, even now, almost twenty years after his untimely death in a car crash in Norwich at age fifty-seven, just as he was achieving true literary fame.

I am teaching my master's class, The Non Fiction Novel, a term I thought I had made up but which I discover has been around since Truman Capote's *In Cold Blood*. The course looks at hybrid forms of fiction-non-fiction, including a currently in-vogue form of auto-fiction in which the writer-author-narrator appears as a character or persona in their own novel. We read Sebald himself for this course, puzzling through the still-life character portraits that make up his work, and studying how Sebald mixes fact and fiction to create his unique brand of anxious histories.

My third coffee of the day cools on the desk. I've been up since 5am. In all the years I've been teaching at UEA I have never moved to Norwich. Instead I keep up my masochistic commute from my flat in north London. I don't mind working in Norwich, but I find the city depressing after a couple of days; it is cold, inward-looking, it has far too many churches. I've never been a provincial city person, but I flourish either in the wilderness or the metropolis, as if they are the same thing.

Sixteen students are ranged around me, cast in the slant of early spring afternoon. Today we are reading Damon Galgut's *In a Strange Room*, a trilogy of novellas, published in 2010 and shortlisted for the

Booker Prize. The book explores the fundamental estrangement of the self from the self. Galgut's thesis is that the self is a construct of memory, as well as a conundrum. Fiction and memory are revealed to be the same entity.

We pause for our mid-seminar break. One of my students, a former trade unionist, approaches. 'Do you think we're doing the right thing here, meeting in person?'

'I don't know,' I say. 'I'm waiting to hear from the university.'

The Coronavirus is now circulating 'in the community' as the government despatches say – which community, I yell back at the screen, the one you destroyed through Brexit and austerity?

When the students return from break I see that many of them do not have coffees in their hands, as they usually do. We have already started to avoid food and drink service stations. The class check their phones all the time. We look at each other as if we have changed into gargoyles overnight. *Is she infected? Will he give me the virus?* The thought flares and dies in my mind, quick as a lit match: *we are the virus.*

At 5pm it is still light. I say goodbye to the students. 'See you next week.' We don't know that this is the last time any of us will be in a classroom, physical version, in a year and a half. In fact, this class will never be together in person again.

Walking back to my office I am overtaken by a vaguely familiar feeling. It is acrid, like the liquid tar used to pave roads. I try to remember where and when I have felt it before. For a while now, my time in Makuleke training to be a walking safari guide, its banishments and its ordeals, has been hovering somewhere on the edge of my memory. I think of those months spent in northernmost South Africa often, probably more than I think of my time gaining my first professional guiding qualification at Lewa, a far more pleasant experience.

Makuleke's ochre boulders, unmoved since the formation of the planet, its nervy kudu and giraffe, their careful gait as they picked their way through its stony tumbledown wilderness, the ochre winter sunsets — what is it about this place that has stalled inside me? Maybe because the rotation of the planet has brought me around to the same time of year as when I was there, March-May, although here in England it is spring, the light is just beginning to return, and in Makuleke the kiln winter light of southern Africa was settling in. I realise that what I am experiencing is not so much of memory, or remembering, but a denser, more forbidding feeling, like brushing up against a nameless but supremely powerful dimension. It has something to do with the rush of history, that feeling of a dominant force overtaking you, just like the fake lioness perched on an electric sledge that we used on the firing range for our target practice, the resoluteness in her sage eyes as she rocketed toward us, how she travelled at only half the speed of a real lion charging, yet she was on us in a snap.

Makuleke, 22.859667°S, 30.8875°E

Lion Charge

May 27ᵗʰ, 2013. Makuleke concession, Kruger National Park, South Africa. 30°C

In Johannesburg we settled in for the seven-hour-long drive to the outer edge of the country. We were headed to the Makuleke reserve in northern Kruger National Park, next to the Limpopo River and the Zimbabwean and Mozambiquan borders. After the long drive, including a pitstop in the closest town of any size to stock up on the usual safari guiding survival kit items (biltong, Duracell batteries, duct tape, energy bars), we finally rattled through Pafuri Gate in Kruger National Park. We tumbled, dusty and stiff, from our minibus and took in the camp that would be our home for two months.

In the northern hemisphere spring of 2013, the year after I passed the course in Lewa Downs, I returned to the African continent, this time to South Africa, where I had lived on and off since 2008. Against my better judgement I was going back to safari boot camp, X-Treme version. Now I was going for the next level up, Trails, the walking safari guide licence.

Walking is the élite of safari guide training. To be on foot in the African wilderness requires a combination of situational awareness, knowledge, judgement and courage. Not that many people pass Trails, I hear on the grapevine. Mostly it is the Advanced Rifle Handling, or ARH, as it is known in guiding circles, that weeds people out. And that's probably a good thing. Because to be a trails guide you take your life and those of your guests in your hands in a way that few other wilderness activities require. If you don't know how to shoot three times as well as your average infantryman, you're in trouble.

I was one of two women in a group of twelve people who were going for the qualification. Most of the students were already working in the industry. The racial dynamics were different from my course at Lewa. The guides from South Africa were white, while those already qualified among us from Botswana and Tanzania were black. Once a preserve of European-descended males, there were now many Africans, and even African women, entering the safari guiding profession. I was classed as an 'international', although I had lived in South Africa for a few years by that point and had a residency permit.

The camp was huddled in a stand of nyala trees. We would live in A-frame tents raised on stilts and have our lectures in an open-air thatched classroom. There was a bush kitchen completely enclosed by mesh to keep out marauding acacia rats and bushbabies. The camp looked comfortable, much more so than our fly camp at Lewa. But there was no fence, and no Kazungu, the camp attendant who lit the charcoal for us at 4.30 in the morning and made our breakfasts amid the lulling savannahs of Lewa.

I studied the map of the reserve, taking in its knit of tangled, melodious names: Hulukulu, Mwambi pan, Reedbuck Vlei, Lala Palm Windmill, Nhlangaluwe, Tsikuyu spring. Makuleke was only two kilometres from the Limpopo River. The Limpopo is the second largest river in southern Africa, after the Zambezi. It switchbacks for nearly 1,800 kilometres, beginning in its birthplace close to Gaborone, Botswana and draining into the Indian Ocean in Mozambique. Even though it was out of sight I sensed the river's commanding presence, its cursive swish through a stand of green-legged fever trees that hem it, unbroken for thirty-two kilometres, making it the longest continuous procession of these trees in all of Africa.

In London, I'd left behind a life of giant Castelvetrano olives, reading Teju Cole, attending Frida Kahlo retrospectives at the Tate Modern, drinking New Zealand Sauvignon Blanc and getting to grips with a relatively new (to me, anyway) image-sharing platform called Instagram. At Makuleke we would begin our days at 4.45am. There was no mobile phone reception. If we wanted internet we'd have to drive for forty minutes to the Internet Rock, a boulder where, if we stood on top of it and waved our phones in the air we might pick up a stray signal from a tower over the border in Zimbabwe. We would eat what the Shangaan women who worked as cooks would make for us, largely meat, meat, and more meat. We would carry backpacks, binoculars, three litres of water, cameras, notebooks, field guides and rifles, which we will have to grip in our hands at all times, as a strap or sling impeded our response time in shouldering and firing. I was carrying fifteen kilograms and weighed just short of fifty.

I tried a joke. 'At least we'll lose some weight.'

Brian overheard me. 'What you lose in fat you'll put on in muscle. Plus there's Violet's cooking. She doesn't like people to go hungry. Most people come out of here heavier than they went in.'

'Thanks for that, Brian.'

If he heard the irony in my voice he didn't let on. Brian introduced us to a forest of rules we must obey while on foot: walk single file, never speak unless spoken to, don't ever leave your rifle unattended, even when you go to pee or shit in the bush; when stationary don't rest its butt on the ground; instead use the toe of your boot to support it; never hold any part of it other than the metal action; when you change your grip, angle the muzzle at the ground.

From the beginning it was clear Brian had read, and probably written, the tough-but-fair instructor manual. On his backpack he wore a badge that attested he had racked up ten thousand kilometres – the same distance as Scotland to the Falkland Islands – in the reserve alone. His eyes were the same taupe colour as the bush. With Brian there would be none of Max's furtive Marlboro Light breaks, helicopter crash videos or sly impressions of a post-apocalyptic Desert Rat.

The next day we piled into the Land Rover and drove to the airstrip, Brian at the wheel. The early morning sun was still low in the sky. It filtered through the marula and timboti trees, the date and lala palms, that clash of highveld and coastal forest unique to Makuleke, casting us in a mesh of crimson shadows as we drove, our faces covered by balaclavas against the chill.

'Don't get used to this thing called a vehicle,' Brian shouted over his shoulder. We all craned forward to hear him, the wind ripping his words to shreds. 'You won't be seeing much of it.'

As we drove I thought, something is wrong with the land. I felt it straight away. What could it be? The Vhembe district, where the Makuleke concession is located, has some of the oldest surface on the earth. We were at the ancient epicentre of the African continent, before it was even Africa. Yet the land was restless. It had antiqueness but lacked the settled quality and confidence of age. I could already hear a sound, but not a singular tone. It seemed to speak in a jumble of voices, all of them talking over each other. There was a clash of systems taking place somewhere. From the beginning I knew in eight weeks we would emerge from the red-eyed cauldron of Makuleke different creatures, and it would not be an ordinary transformation.

'No room for *dwallies*, you *mampara*. Anyone who drops a round is off my range!'

Brian strides toward us, an expired cartridge gripped in his fingers, calling us idiots and layabouts in Afrikaans and Zulu respectively.

Our rifles lowered, we bend to the ground and snatch up our *doppies*. Brian makes us collect expired cartridges off the ground straight away after we have emptied the chamber. This is another piece of South African army discipline now so inscribed in us it becomes muscle memory. Brian drills us on every action to do with shooting – shouldering, finding our sightline, taking the shot, chambering, firing again, picking up *doppies*, his old-fashioned track-and-field-coach stopwatch in his palm, until we can vacate a chamber and slot in our rounds in ten seconds flat.

Sweat trickles into my right eye, stinging it. I need to wipe it away so I can see through the mechanical sights on the end of the rifle – no binocular scopes for us – but my finger is on the trigger now and I am committed to taking the shot. I know what Brian will say, if he catches me. *Bloody well put up with sweat in your eye and take the shot – you think the lion is going to wait to kill you while you fix it?*

The Winchester Magnum .458 sits like a piece of lead on my shoulder. This is the heaviest and most powerful, in terms of muzzle velocity, of our rifles, and shot for shot one of the most powerful rifles on earth. For some reason I find it more comfortable to shoot than the .375, which has less thrust and so ought to be easier to handle.

Today we are practising the simulated lion charge, the hardest exercise in our Advanced Rifle Handling assessment. We not only have to get a shot in at ten metres or less but it must be a brain shot, delivered to a tennis-ball sized area above the left eye. Brain shots are the only way we can stop animals of the size of elephant, rhino or buffalo, or lion for that matter; if you don't disable their nervous system immediately the animal will simply keep on coming.

There is a choreographed sequence of movements we have to produce once the charge is underway: shout to the guests to get behind you, shoulder, fire, check that the 'lioness' is really dead, shoo your clients behind you again, declare the situation under control (hopefully), then test the 'lioness' by prodding her with your rifle. All this needs to happen in thirty-five seconds.

I walk back into position down range, rifle in my right hand, my grip on its metal action. At the turnaround point I break into a jog just as the lioness rockets toward me on her haunches. Fortunately, she is only a poster propelled by an electric sledge at half a lion's actual top speed of eighty-eight kilometres an hour. Still, it is hard to even get into position to take the shot, never mind land it. What hope do we have when faced with the real thing?

I get into position, shoulder, fire in good time. I get a shot in, but it lands in her neck. Normally my aim is good but with practice my performance on the lion charge exercise gets worse, not better. The issue is my positioning; it comes too late and too close. I transfer my rifle to my left hand and walk back to try again. The mouths of my fellow students, who are watching from a nearby ridge in the shade of a tree, tighten. In situations like these I draw on an old tenacity. I will do it over and over, until I get it right. But Brian has other ideas.

'Okay,' Brian dips one lobe out of his ear protectors. 'Let's call it a day.'

We tuck our rifles into their hit-man duffel bags and climb in the open Land Rover. We are too tired to talk so we take in the view: the green snake of the Limpopo, its banks dotted with olive islets of baobab, white syringa, jackalberry.

Brian brakes suddenly at the foot of a nyala tree. At first we can't see what has caught his attention. Then we spot a dark ruffled form. A Verreaux's eagle-owl, the largest in this family of birds on the African continent, regards us with a schoolmasterish glint. The braided cords of his legs are wrapped around a branch. The owl blinks and we are treated to his dusky pink eyelids. We have startled him. I feel a grain of shame. However hard we try to make ourselves ghosts in the bush, we are constantly interrupting the lives of animals.

The Lowveld sun drops over the lip of the land. The owl's eyes open in a flash, as if he has received a signal. He swivels back to us, huddled underneath him, our necks craned. His gaze lands on me. For a thrilling moment we do not look at but see each other. Then he unfurls his wings and erupts into the coming night.

The Ivory Trail

A tall, curly haired man peers at us from underneath a floppy bush hat. With the muzzle of his rifle he draws a circle in the sand.

'What can you tell me about this elephant?'

We squat on our haunches in the dust, staring at dinner-plate-sized imprints. 'It's a sub-adult, probably two years old,' we say. 'It was travelling north yesterday, probably in the evening, heading to drink in the river.'

'Good, guys, good,' the curly-haired man says. 'Now let's go find them.'

Adam appeared halfway through the course, a wild-haired, mild-voiced man, twice as tall as his good friend Brian, his knock-kneed stance and sandalled feet making him look far more delicate than the standard issue safari guide. Adam has a particular role to play in our training. In order to attempt our trails guide qualification we have to rack up a certain number of *encounters* – this is the official term for stumbling upon a herd of buffalo submerged in a sea of grass, or meeting a lioness sashaying down a riverbed in search of the wilderness equivalent of a takeaway. It becomes a word we use many times every day, bleached of its sexual innuendo, more evasive than *meeting* but less alarming than *collision*. If we don't naturally encounter enough encounters for our Dangerous Game Logbooks, Adam will arrange them.

As you might expect of someone who is a matchmaker with mortal consequences, Adam is extremely calm. His voice is quiet, hushed, even yogic. 'Keep present. If you allow your thoughts to drift you get surprised, which means you surprise the animals. And what happens when wild animals get surprised?' He doesn't wait for an answer.

In the early morning cold Adam takes us swiftly along the Limpopo under a canopy of white-green fever trees. We walk narrow trails carved by zebra and eland through swamps where crocodiles masquerade as logs. As the hours pass, heat congeals and the gold axe of the sun cleaves our heads. He tells us that the successful bush walk is the one where the animals do not even realise we are here, that the trails guide chaperones clients around the bush like a Greek chorus, silent observers in a spectacle of devoural, decay and survival.

Suddenly, he makes the *shhh* gesture. We cascade to a domino-like halt that would be comic in other circumstances. Adam gives us the most marginal frown, but enough that we are chastised.

Out of the stillness of the morning, a sound emerges. A swishing, pleasant noise. As it turns out, the sound of twenty kilograms of wisteria being masticated is soothing. Ahead of us, not ten metres away, an elephant bull's trunk and ears swing in a wide, pleasured ellipse. None of us – including Adam, who knows a thing or two about elephant – had seen the animal until we were up against it. That a two-tonne truck of a creature can be virtually invisible is one of the elephant's many talents.

We stand and wait. The elephant continues, unaware of his spectators. But if the wind were to switch direction, this is what would happen: his ears would become rigid, his left foot would come off the ground. He would swivel his head in our direction and hold his trunk aloft like a periscope. After that, anyone's guess. Worst case scenario: a charge that would scatter us into sharp thornbushes and possibly grind slow-moving outliers, such as small women carrying too much weight in the form of a rifle and backpack, into an ugly pulp.

The elephant keeps on chewing, his huge yet humanoid bottom swaying happily. It is hard to describe the feeling of being close to these animals, close enough to touch their bristly, puckered skin, to look into their garnet eyes. Their intelligence radiates outward, a signal, a force field. Their mystique comes not from their size, or not only, but from the aura they emanate, a palpable zone of cleverness and alert. Their intelligence twinned with their bulk feels like being confronted with a deity. When they run, they are largely silent, thanks to a network of cushioning cartilage and over 100,000 capillaries in their feet. An onrushing elephant can paralyse the most experienced guides. Frozen in place, they fail to take a shot, or run and so they die.

After a few minutes of basking in the elephant's quiet charisma, we file away. 'We got away with it,' Adam grins. At a safe distance we rest in the shade of a marula tree. On his phone Adam shows us a video of a few years previously in Botswana. In it, a monumental bull emerges from a stand of trees, only ten or twenty metres away from the camera. The bull stops in his tracks and stares, his tusks flanking him like scimitars.

The camera dips suddenly to the ground. Then it flicks up. Now we are looking up at the trunk, held aloft, searching the air. The graphite body of the elephant towers above the camera like an aircraft carrier.

'Has he flattened you?' we ask.

'No, I lay down.'

'You *lay down*?'

'Yeah.'

'Then what happened?'

'He walked away.'

In Botswana, where Adam qualified, guides are forbidden by law to carry firearms. Instead they carry umbrellas and a roll of double-pleated toilet paper. 'You open the umbrella when the elephant starts its charge, then you grab hold of the end of the toilet paper and chuck the roll at them.' Adam throws his arms out in front of him like a sleepwalker. 'The umbrella fazes them because it's a sudden obstacle. And they think the toilet roll is a big snake. Plus elephant don't like the colour white.'

We frown.

'It works, guys,' Adam says. 'Trust me. I'm still alive, aren't I?'

Makuleke is on the Ivory Trail, a killing corridor that yielded hundreds of tonnes of ivory in the first phase of the war on elephant at the turn of the twentieth century. The animals had been hunted systematically on the very land we were training on, along the Limpopo. 'The elephant here remember,' Adam says.

'Do you mean as in, intergenerational trauma?' I ask, expecting him to fob me off.

'Exactly that. The elephant here are the most unpredictable I've ever encountered in Africa. They are not at peace. Two guides have had to shoot to kill in just this past year.' He shakes his head.

Makuleke had a long human history. Just over a thousand years ago, a gold-trading kingdom reigned over the area. Thulamela was a thriving metropolis, perched high on a hill between the confluence of the Limpopo and Luvuvhu rivers, only twelve kilometres as the bushshrike files from our camp. Then, a thousand people, many of them artisans, forging gold, carving ivory and stone, lived in the citadel above while another few thousand are thought to have lived on the plains below.

Thulamela flourished for half a millennium as a trading centre between the inland civilisations of the Low and Highvelds and the coastal trading cities of the Indian Ocean. No-one knows why it declined, and nothing of it was known until 1983 when a Kruger Park ranger stumbled on fragments of ceramics and the remnants of dwellings, deep in the bush. Since the demise of Thulamela, the Makuleke people lived a subsistence existence on the jumble of hills and rivers. In 1967, the Apartheid government evicted them and forcibly resettled communities in the Bantustans, the areas it had gazetted for 'Blacks', a system similar to First Nations reservations in North America.

One day, while out walking with Adam, and following a group of white-throated bee-eaters, we happened upon the remains of the Makuleke people's hillside village. Only the skeleton of one hut remained. On the ground were twisted spoons and fragments of pottery, bleached and distorted by the sun. There was a desolation to the place I'd never before felt in the wilderness. The remnants of the village were enough to imagine a community: children playing with flocks of spotted guineafowl, shooing them between tight-fitting mopane, women bent over charcoal fires.

We stopped and rested in the shade. The day was hot, for winter: thirty degrees. Adam took off his hat and wiped his brow. Even in the depths of the African bush, Adam told us, there was a human history. 'You don't exist for one-and-a-half million years – and that's the modest estimate, it's probably far longer – without something remaining behind. But only a few hundred years after the fall of Thulamela, how many of us would be able to survive here, if we were dumped without food and water and only the clothes we stand in?'

It was a rhetorical question, one all my guiding instructors asked of us in various ways. What has happened to us? We are now effete creatures, sieved of our animal instincts with our stock portfolios and catalytic converters and turmeric smoothies, yet more defenceless in this realm than an impala. *Situational awareness, situational awareness* – Adam drilled the phrase into us like a mantra. What he meant was that we had to adopt a fusion of hyper-awareness, confidence and knowledge to counter our disadvantages, which we are unused to having. He was talking about an alertness to the signs and symbols of the bush, rather than the strange voltage of the human. When you vault yourself out of the human world for a long period, you begin to perceive how animals experience us: the instable current that wafts from the human, and which is a mix of brain waves, bodily frequency and something you might call soul-energy. We had to be aware at all times of the wind direction, Adam said, the birds you hear, the smell of the land – 'and if you get a bad feeling, stop.'

'What kind of bad feeling?' I asked.

'You'll know it when it comes.'

'But I feel it all the time, here,' I said.

'What do you feel, exactly?'

'I don't know. That something is wrong.'

Adam stopped and regarded me neutrally. 'And you think you know this land? How long have you been here?'

My fellow students looked apprehensive. Until now, Adam had been jokey and comradely. Now a chill coated his voice. I had a sense of what it would feel like to receive his disdain.

Everyone waited. I said something non-committal – it was just an intuition, I was new in this landscape, I was probably wrong.

But I was not mistaken, I knew. Makuleke was the most dangerous wild place I'd ever spent time in, apart from the Antarctic. I thought of the photo of the endless black mamba that had been captured a couple of hundred metres from our camp a few years before. The elephant were judgemental and wary, as Adam had confirmed, and the Kruger Park's rhino were under attack from poachers from Mozambique. It was quite possible we would stumble upon these men one day and, they, seeing we were armed, would engage us in a gun fight. Almost every night I heard lion close to my tent. In the mornings when I was on breakfast duty, I walked to the kitchen knowing they were probably only metres away. But the feeling of unease the land broadcast was more than the sum of any of these dangers.

That evening Adam delivered us back to camp from another epic walk. We had been on our feet for at least ten hours. I shucked off my walking books and sat on my veranda, drinking a cold Appletiser. It was late May and winter was deepening. The waning afternoon light was the equivalent of lens flare and cast the land in a bronzed nostalgia. It ought to have been pretty, but I was overtaken by a sense of enclosure.

The others awaited around the fire. I should head there before the night gathered, yet I stayed on my *stoep*.

Not for the first time, I was struck by a paradox at the heart of my quest to understand the wilderness. There I was in one of the most pristine wild landscapes in Africa, and to know it intimately I had marooned myself with a group of people with whom I had nothing in common, apart from a pull toward the African bush. What I wanted most of all was to be alone in vast landscapes, to dissolve myself into their seigneurial spirit. But to even get close to achieving this required me to sign up for a wilderness version of Big Brother.

The sun dissolves into a cold amber sky. The winter night in Makuleke is thin and cold. The trees are iron under a gruff moon. I think, in this land we will all turn sombre and severe, like Brian. Maybe he had always been so, or perhaps he was once pliant and yielding. I envy how he has only found his soulmate in this place, the physical embodiment of his inner landscape. Now they are married to each other, forever.

The Firing Range

And again! And again! Shoulder, sights, fire. Shoulder, sights, fire.

We are back on the range, shooting fifty rounds a session. I absorb the recoil better than I expect, but I am not a pretty shooter. Compared to the hale Highveld youngsters who tuck the rifles their arms as easily as if slinging a lunchbox over their shoulders, I look like a nineteenth century musketeer crossed with a Merce Cunningham dancer. Afterward, my body buzzes from the detonation of five thousand feet per second muzzle velocity. It feels convulsed but also enlivened, as if I've developed a sudden amphetamine addiction.

The sun rotates above our head, a gold pendulum. From somewhere in the riparian forest near the range comes the fluting metallic call of the tropical boubou. He sings on repeat, his whistle ribboning over the Limpopo.

In the distance, bodies move through the forest. Over the past few days a steady stream of Zimbabweans have crossed the shallow river on foot, walking unimpeded by police into South Africa. From time to time one of us is despatched to warn them away from our firing range, shouting that our bullets, if they escape the banks of the range, can carry on for two kilometres. They barely take any notice of us; they will be walking for at least a day among lion and leopard and elephant, unprotected, and we are not high up on their danger list.

I lower my rifle. Brian approaches me. I steel myself. What am I doing wrong now?

'You've handled rifles before, haven't you.' It is not quite a question.

'Yes.'

'Well I want you to forget anything you ever learnt from anyone else. Your aim is not bad, but you haven't got a scope now, so you have to get your aim right on the *mik*,' – he uses the Afrikaans word for the manual metal sights. 'Not once, but over and over again.'

He starts to walk away from me but keeps talking, directing his words at the air between us and the target, which shows a purple-nosed buffalo. 'Without thinking. Without even trying,' Brian goes on. 'Until you can do that, you're not getting off this range. Understood?'

I am ten years old. My great-grandmother, who is by far the best shot in the family, shoves a .22 into my hands. It is she who teaches me, as she had my grandfather before me, to shoot rabbit and ptarmigan. We hunt not for sport but to survive, just as our ancestors had done, and before them the indigenous peoples who taught our ancestors how to stay alive in this unfamiliar land.

I fake being a bad shot, deliberately missing the white rabbits that skitter across the snow surface. I am not ready to kill anything; I already have a saviour complex.

'Good enough only for the kitchen, are you?' she crows.

My great-grandmother is a multi-talented backwoodswoman with a wide mean streak. She stands me on the bed so I can see the boxes of bullets that we keep above the wardrobe in my bedroom. I read her what is printed on them: 30.06, .22, .303. I wonder if the fact I learnt my numbers from bullet boxes has led me, somehow, to Makuleke.

Fast forward to a run-down apartment on the edge of the North American continent. Thirty metres away the Atlantic Ocean ploughs back and forth, back and forth, with that plasmic tenacity only oceans possess.

We stay in this apartment for a time after fleeing the breakdown of my grandparents' marriage, partly precipitated by my great-grandmother selling the house we lived in without telling us. After pulling that trick, she stays behind in Cape Breton, snug in a comfortable retirement home on the proceeds of the sale of the house we thought was our home.

Now I live with my grandmother. The extended community of cousins, sisters, aunts and brothers has disappeared. We are on our own in a city where there are ferries, strip malls, drug stores, traffic lights, and two gigantic thread-thin bridges vaulting over a deep harbour.

A cold fog hems the city. My grandmother sits at the kitchen table, smoking and playing a game of solitaire. An abstract panic overtakes me. I begin shaking, very slightly, as if I have a chill.

'I have to go.'

'Go where?' My grandmother casts around the apartment. 'You mean the corner store?'

'Somewhere,' I say. I mean, anywhere. Where can I go? The corner store that sold Baxter's chocolate milk and Love Hearts, the ferry terminal at the bottom of the hill, the four-lane highway whose roar was audible in the apartment, lined with Zellers and Dairy Queens?

I have a premonition. We are poor, and getting poorer. I am going nowhere, literally and metaphorically. I am being lined up by circumstance to serve as one of life's bystanders. Rightly or wrongly I conclude that I can dodge this fate through mere motion.

Meanwhile winters in our land are still a force to behold. Each year in October the world ends in a slow white apocalypse. Cowed by the onslaught, the land falls into a cryogenic slumber. The cold is so total it freezes our feelings dead in their tracks. Codfish throng the oceans – they really do *throng*, the sea so thick with them it is a matted blanket. Few planes unzip the skies above our province on the rim of the Atlantic Ocean.

The world feels empty, still. The planet's habitual solitude is appreciable in the long mute minutes I spend staring at the steel ocean, dull interregnums that last a lifetime. What I do not know is that carbon from distant mythical eras with names like Silurian is being dug up and burned and put into the atmosphere and that the planet is warming, so quickly, under our noses. I don't know that I am alive between an eternal pas de deux between ice ages that will never recur, that our winter world is ending.

'You need to know that when it comes to it, and only then, you will take the shot,' Brian tells us, until it is a mantra in our heads.

I heft the rifle on my shoulder and aim. A wet-nosed Cape buffalo stares at me. I imagine the buffalo coming to life, leaping off the poster and charging pell-mell. The oxpeckers that had been flensing its flanks of ticks rocket into the air, the clump of dried mud stuck to the buffalo's horns is flicked into powder. I feel the thud of the earth as hooves consume the distance between us.

I could perform the same manoeuvre as I used at ten years old, so that a rabbit might live another day. But a rabbit will not kill me and a Cape buffalo almost certainly will. The muscle memory that Brian is inscribing into us in these hours on the range when the sun falls mitre-like on our heads will be ignited. I will pick up the rifle. I will shoulder and fire, killing the buffalo with a single shot to the brain. I know I will do this. I am resolved.

Black Mamba

'All right, we know the rules. Walk in single file, turn off mobile phones and camera shutter noises. Stay close to me at all times. And whatever happens, don't run.'

The day of my practical assessment has come. We form a circle in the draining heat of the late afternoon. My fellow students will impersonate my safari 'guests'. Briefing done, I turn and walk us out of camp and through a glade of wild date and lala palms.

As soon as we set out, I am aware of how much I have had a follow-the-leader mentality in all our training walks. We all feel safe – or as safe as you can be in the bush – with Brian and Adam, who have fifty years of guiding experience between them.

Now that I am out in front, with no thick-calved man between us and whatever the wilderness might throw in our direction, my back stiffens. My eyes sharpen, like pencils being whittled to a point. A flashing neon sign erects itself behind my eyes: PAY ATTENTION.

I lead us out through the salt pan to the north of our camp, a place we have walked through countless times. I notice the bulk of the wild date palms near the windmill, how their thick fans are more than capable of hiding an elephant. I register an alarm, just before Brian and Adam move up to stand a few steps behind me. I quickly glance back at them. Their grips on the rifles have tightened.

'Let's stop here,' I say. There is no game to see, but to gain time to assess the situation I recite a spiel we all know now about date palms, how palm vultures will land in them on their commutes to the coastal plains of Mozambique, how the vultures, once common, are now a rarity, how you can drink the wine made from the wild date palm if you have a strong stomach.

The palms are taller than I remember. Why have I got it wrong? Now that I am responsible for everyone's lives, the world has grown larger.

The sun slices my eyes. Brian has discouraged us from looking at our watches; we must tell the time by the slant of the light, and at night, the stars. This is one thing I am good at – I can reliably tell the time within five minutes. *Four thirty-five*. I have to find another way or it will be dark soon. To keep a group out on a bushwalk more than fifteen minutes after sunset is an automatic fail.

A path presents itself through the date palms. If I skirt round to the east, I have good visibility. I will have time to perceive the shadow that turns into an elephant. Then I need to get us through the fever tree forest. I catch sight of a path through the tall grass. Here dangers also lurk – buffalo, those free radicals of the bush, ricocheting around without warning or logic.

'Let's have a look for some eland in the fever tree forest,' I say, forcing a lightness into my voice. Just as I begin to wade through the tall sage-coloured grass, I hear a rustle to our left. A family of black-backed jackal rocket out of the grass, where they had been resting. Everyone laughs nervously. Jittery guests, even if they are my fellow students, are not a good sign. I am failing to instil confidence in them. Which is not surprising because I don't feel confidence in myself, either.

As we swish through the grass Brian and Adam fall behind, a sign I have made the right call. I lead the group along the route I have chosen. *Oxpeckers? Buffalo or rhino could be nearby. Visibility of terrain? Good – open plains, no surprises. Wind direction: southwest. The hippo won't smell us.* My flâneur brain rebels. It wants to just walk, admire the view, think private thoughts. It is not impressed at having to give over all its computing capacity to survival.

Two hours later we reach our endpoint. In the interim I manage to skirt us safely around a herd of buffalo and pass three hippopotamus in the pool in the late afternoon, a time when they are often returning to the water to feed and particularly dangerous.

In between these encounters were moments of spectral beauty I couldn't relax enough to appreciate, such as when we moved among a mixed herd of eland and zebra in the fever tree forest in a golden haze of late-afternoon lucre. There, we listened to the sharp flute of the tawny-flanked prinia and the trill of the fiery-necked nightjar. When I cited their names Brian gave me an approving look. At the Land Rover I serve cool drinks and crisps to the rest of the group. That night I fall asleep at 7.30.

The following morning Brian gives me my result.

'I was going to fail you, but you recovered yourself well, so I'm passing you with a sixty-four. You'll have to make up your marks on your theory test and rifle handling.'

My mistake, Brian told me, was to lead the group too close to the wild date and lala palms. My error was serious, but because I had stopped before we reached the likely range of a surprised elephant and re-considered, I would be reprieved.

'You should have skirted them and headed straight for the grass. But you did a good job with the buffalo. We saw more dangerous game on your walk than all the others and by and large you made good calls. Let this be a lesson. You were doing too much thinking. Guiding is about *looking*. Plus you know your birds.'

Thinking, looking, doing. In the bush, these become separate activities. In my other, normal, life, I take them for an alloy of each other. Like the cheetah who relies too much on mere speed for hunting, I have become over-specialised for thinking, and this is not what is required here, in this wild place. Thinking can get the people who have placed their lives in your hands killed. I have made not a procedural but a moral error.

I retreat to the *stoep* of my tent with its enigmatic baobab. I feel like crying, something I haven't done in a very long time. I used to be a crier. My eyes would tear at the sight of a pigeon that had lost its foot or even a vaguely sentimental film. But now I can't remember the last time I cried and to give in to that convulsion in this unforgiving place feels aberrant, more damning than shame.

My life has taken a strange turn, I consider. I feel at home in the African wilderness. It speaks to me in a way my native landscape never did. No doubt it has something to do with eternity, and with the lure of dissolution into a complex web of associations, perfected over aeons. I had begun training to be a safari guide after the walk I'd done across the Namib desert two years previously, because I wanted to know the African bush and be capable in it, to the point that no-one – maybe I was thinking no *man* – would have tell me what to do to survive in it, ever again. I had begun training to be a safari guide for the bush knowledge it would give me, not to actually qualify. But somewhere along the way I'd committed to following through. Now I was considering doing it for real, professionally, for a living. For the first time since I had begun to think of myself as a writer, I had embarked on something I did not think I would write about, or rather I was not doing these professional qualifications in order to write about the experience. In any case, I had run out of steam with writing. The issue was not writer's block, I had no shortage of ideas or drive, but increasingly I couldn't stomach the aspects I think of as externals: the publishing business, success, prizes, money.

By that point I had published six books of fiction and I was working on what would eventually be published as *Ice Diaries*, my account of being writer-in-residence in Antarctica. All I wanted to write about was landscape and nature, but I could not find a way to make this fit with our cultural obsession with story, as in the narrative arc that is recycled through our culture as if caught in a permanent washing machine cycle: the mystery of the self, overcoming challenge/trauma, ambition, fulfilment, enlightenment, transcendence. Surely there was another story that we were missing, about the accident of us, our ephemerality, of what the earth would have experienced had we never evolved?

My growing awareness of my limitations as a writer had gone hand in hand with my increasing interest in nature and the wilderness. Spending long periods of time in one slice of wilderness leads to a state I have only felt while writing, a kind of hypnosis built of a fascination with pattern. More than anything else I had done in my life by that point, training to be a safari guide required me to define who I was. Like a mirror, the experience stared back implacably at me. *Who are you?*

Who am I, really? I am a welfare-class woman who grew up in a trailer on the side of the Trans-Canada Highway on a periphery of Canada's already peripheral east coast. Could I be putting myself through this ordeal simply because I was attracted to the irony of someone like me, or actually me, rumbling around in tank-like Land Rovers and knowing how to shoot a hippopotamus dead with a single brain shot while people from San Diego dressed head to toe in Ralph Lauren who make more money in a day or week or month than I ever will in a lifetime cower behind me? Perhaps becoming a safari guide was just an elaborate form of class revenge.

The bronze-winged courser pipes up. This bird strikes up its marimba tune at the same precise moment every day, 5.58pm. Its call and regularity both are soothing, as if there really is a code under-writing all reality, and I have stumbled across it.

Darkness descends fast, as it always does in Makuleke, in a single sheet of velvet. Before we light a fire we tidy the plastic chairs in the lecture theatre. We stack them at the top of the stairs to deter hyena from entering under the cover of darkness (hyena dislike chairs). As we assemble around the fire night dissolves like a collapsing frieze, a fissure in the skin of time. I am an anxious type normally, yet in this lethal realm I feel alarm or alertness, but rarely fear. This is one reason why I think I am made for this place, the African bush, despite having begun life twenty thousand kilometres away on the edge of the Atlantic Ocean, on a distant latitude where Greenland-born icebergs went to die.

On one of our final days in Makuleke we set out in the vehicle to our departure point for our walk. Brian is at the wheel. He nudges us through the bush. That day I am the tracker, which means I am perched in a seat bolted onto the Land Rover's bonnet from where I can see the road ahead.

Although I have the advantage in perspective, Brian has the experienced bush eyes. We spot it in the same instant. From afar it looks like a grey ribbon tying itself around a bush: bows, serpentines, figure-of-eight knots. Its length is an instant field characteristic.

The black mamba is amongst the most venomous snakes on earth. A bite will kill a human in under thirty minutes. It is also one of the most aggressive, indestructible snakes: they have been known to launch themselves at safari vehicles and, in 2016 in South Africa, an actually dead mamba unwisely handled by a child on a safari convulsed and managed to land a bite (the child survived, but only just). Just before we started our course in Makuleke a single mamba had killed every member of a family as they lay sleeping, eight in all, when it penetrated their shack in KwaZulu-Natal.

Brian slows the vehicle but keeps going forward. He brings us to a stop some distance away, around twenty metres – a safe distance, technically. The snake is very long, as many mambas are, perhaps three metres. We have disturbed it. I watch it unwrap itself from the bush. I am amazed by its speed and sinuosity.

Something drains out of me. Inside me the liver is sliding over to where the kidneys should be. My kidneys meanwhile float upward to become my lungs. For a second I am entranced by this weird sensation, and forget to be afraid.

Suddenly, the mamba sinks through the bush nose down. When it hits the ground it appears to explode. We watch it disappear, a terrestrial flash of lightning, into the grass.

I glance back into the vehicle. Harald, an already-qualified African safari guide from Tanzania and Wouter, a similarly experienced bush guide from Durban, neither of them shrinking violets, look at me, their faces a strange green-grey; the same colour, come to think of it, as the snake itself. (Mambas are not actually black, but an olive-silver; the name comes from the colour of the mouth of the snake, which – if you are unlucky enough to ever see it – is a dark glittering charcoal.)

'We thought you were going to bolt,' Brian says, from behind the wheel. An echo flashes through my mind: The Bolter in Nancy Mitford's *The Pursuit of Love*, the mother who abandoned her baby to live with a succession of men, and who fled at any sign of responsibility or trouble.

'I'm not a bolter,' I say. It's true; if anything my flaw is to have too much stamina and tenacity. Sometimes, you just have to give up, or giving up is the best thing you can do. But I often fail to read the runes and stay, out of sheer will.

We drive on. I stay in the tracker's seat, an arid wind against my face. For the rest of the drive we are unusually quiet.

Death is all around in the bush, I think. That is just the fact of it. Every day we see an animal kill another animal, or find evidence of recent animal-o-cide. Each animal in Makuleke faces a black mamba or a version of it, day in, day out. Their power is the same as their vulnerability; that is what makes them animals.

For most of human history animals have been magical beings. We have made of them entertainment, neutered companions. Animals oversaw our species' transition from nature to culture, providing us with deities and pinwheel systems such as the Zodiac. Now we are disappearing them. The land is already freighted with their absence. Perhaps this is the reedy, empty tone I pick up in the land at Makuleke:

resentment, twinned with the same ghostliness I felt crossing the Atlantic Ocean on the *JCR* three years previously.

As we drive I think of the conversation Max and I had around the fire at Lewa, about why people who spend too much time in the African wilderness become corroded. They are attracted to the bush for its elemental simplicity, he said, they think they will be enlivened, stiffened into a rectitude that organised, safe twenty-first century life robs us of. So many guides and a fair number of tourists spout this false belief, that nature is more real and more honourable, and such conviction confers a nobility on those who harbour it. It's sanctimonious, a false moral hauteur. In reality the bush is an eternal tragedy worthy of Euripides, a play titled *Life and Death – but Mostly Death.*

We return to camp in the dark. As we clamber out of the Land Rover, Wouter gives me a wry look. 'So what did you make of death-on-a-stick?' He means the mamba.

'I don't know. I wasn't afraid. Or, it didn't feel like fear.' What I do not say to him is that I felt something brush up against me as we watched the snake, not an object or feeling but a dimension.

'There's nothing wrong with fear,' Wouter says. 'That's your brain keeping you alive. The internationals like you who come here and walk around barefoot and think they're in bush paradise. Well –' he purses his lips. 'You know what happens. Fear is your friend. When it comes calling, don't ever send it away.'

At the end of the course we were driven the seven long hours back to Johannesburg, ascending from the Lowveld to the Highveld in the bright thin glare of the southern hemisphere winter. Suddenly we found ourselves back in the world of 3G signal, petrol stations and restaurants where you could choose your own food.

My fellow students had entry-level jobs as Third Guide lined up at five-star reserves and resorts in Mpumalanga or the Kalahari. I was going back to supervise creative writing master's dissertations under the taut skies of a Norfolk spring. By nightfall I would be accepting the breezy smiles of a British Airways crew who would convey me to London, where there were no baobabs or bronze-winged coursers to sing in the evening.

The bush receded, even if the memory of Makuleke's laterite corridors persisted inside me, a simmering ember. I knew that no matter how hard I tried to hang on to it, my bush knowledge would dissolve. Very soon I would not be able to ID at a glance fifty species of tree and bush, both Linnaean and common names; the grazing uses of twenty types of savannah grass; two hundred and fifty bird species identified with only a broken bar of their call; how to track a kudu, whether it is injured or fit; the frequency of elephant communication in hertz; the gestation period of giraffe; why wildebeest migrate; all the antelope and characteristics, Sharpe's grysbok, bontebok and steenbok (there are a lot of boks); how to line up a kill shot for a charging elephant; what vultures in the leadwood tree mean; how to clean a rifle from end to end in ten minutes flat blindfolded; how to navigate by starlight a labyrinth of young mopane as densely disorienting as any Tudor maze.

We had to say goodbye. We students had been through a lot together. I thought again about the effects of spending so much time – eating, studying, walking – with a small group of people, how, when they are gone their voices remain for days or weeks, banging around your head like spirits trying to find a way out. I can still hear Adam's voice – level, assured, almost on another frequency than the human – saying, *I am an African, I have lived here all my life, but I still feel a thrill when I hear the word 'Africa'.* I was partial to that romance. I was still in thrall to the bush; its hold would not lessen, and maybe never will.

But I was dogged by another story. Let's accelerate one, two, ten years from now. See a woman at the wheel of a Land Rover rumbling around a reserve, earning five thousand rand a month after food and board. Watch her drive honeymoon couples from Durban or dentists from Chicago who command her to find a leopard and who shout and whoop at elephant, scaring them away. This woman lacks the natural authority – they would listen to a man, especially a South African man – to stop them.

See her ricochet from one lodge job to another in private reserves owned by diamond magnates or hedge funds registered in Bermuda, one of an army of low-paid workers, the supply of which is endlessly replenished with new stock, often knock-kneed, braai-fed young-sters from Highveld farms that are encircled by violent shantytowns inhabited by men and women who will always be landless. Watch her drift into late middle age, earning less and less as she loses whatever currency she might have had to fresh-faced guides with names like Angel and Lance, still *bush-bevok* ('bush-mad' in Afrikaans), trapped in a diminishing obsession.

PART VII

THE LAND OF
LETTING GO

April 24th, 2020. London, UK. 18°C

Sharp rainstorms drift above London then sail on, leaving layers of ivory cloud, light seeping between their seams. The months-long spring of the British Isles is entering its terminal phase. The skies harbour a new, scrubbed, note.

We entered so-called lockdown thirty-one days ago, on March 23rd. *Lockdown.* I don't like the hollow zeal in the word, previously heard only in military circles or countries under martial law. I am shocked how everyone rushes to use it, as if they can't wait to turn the key to their own cells.

The days are long and undifferentiated. The sound of my own voice booms in my head, as if I am living in a ballroom. *You have to sort out your finances. You might not have a job in a few months' time and the income from your books certainly isn't going to save you, is it?* The Overlord is back, the me-and-not-me voice.

A few weeks into our confinement I have a strange experience. I undertake an expedition from my desk to the coffee maker – this is what now counts as an event. On the way I am seized by a vivid yet abstract feeling, a bit like déjà vu, but with an echo of panic. Suddenly I realise why the unfolding pandemic feels so familiar.

In 2009 I published a novel, *The Ice Lovers*. It was about the Antarctic, ostensibly, but like all novels it is never 'about' what publishers and reviewers say; fiction's about-ness is only ever a peg off which to hang a complex costume. In one of the sub-plots, a year or so before the main action takes place, a pandemic had swept the world, trapping one of my protagonists in her London flat for months. Until that moment on the way to the coffee machine I had completely forgotten about this storyline in my novel.

I pluck the novel from my shelf. People – characters – I had conceived and then forgotten about spring alive from its pages, like one of those three dimensional friezes in Christmas cards. Helen is a journalist. From her London flat she watches as a virus lays waste to humanity and normality outside her window. She observes the dog walkers: 'I saw them, sometimes, in the wedge of gritty green that passes for a park... they wore white masks and looked around them, constantly, as if they could spot it coming.' She notes the 'stiff choreography' of people avoiding each other 'like dancers in a Merce Cunningham piece'. Airports are shut, and – just as I still look futilely into the sky for planes threading through the waypoints to Heathrow – she finds she 'missed that sound, of hydraulics, the groan as the plane's bellies slit open and the landing gear dropped down, more than I ever would have believed.'

After the novel was published, I regretted the pandemic scenario – it was too claustrophobic and terrifying for readers to engage with. I did no research or background reading on pandemics before writing the book because I could not face the dire-ness of the scenario so I simply made it all up. At the time, my reasoning for it was that I needed my characters to have been through a collective experience which profoundly changed their relationships, not only with themselves, but with the world, before going to the Antarctic. I wanted them to be familiar with the experience of having what you thought was reality completely recast before your eyes. Also, I needed my personnel to arrive on the continent wounded, in disarray. The Antarctic would reconstruct their faith in the world.

There was another theme I wanted to explore in the novel, but I was wary of writing about it. In Antarctica, for the first time since I had been a child, I heard or perceived or felt not a presence as much as a voice. The sound of other consciousnesses in your mind is associated with several mental disorders, including the nuclear bomb of all dissociative conditions, schizophrenia. The last thing I wanted to hear, ever, was someone or something else talking to me, rattling around the private domain of my own mind.

My characters also heard this voice, emanating from the continent. Beyond broadcasting a voice, Antarctica had a unique power, one I reflected on in the novel while failing to understand its effect in my own life. Antarctica was not a place, or not only, but a dimension. Once my characters knew the continent, they mutated into different people. Antarctica had a transformative effect on them and on me, although it would take me years after leaving the continent to identify its true nature: a possession.

Rothera, 67.5678°S, 68.1267°W

Bluefields

It was an ordinary morning on base. We all drifted into the mess hall to make inadequate cups of coffee. Clumps of Nido, a form of powdered milk which gave everyone constipation, floated in our mugs as we arranged ourselves at the long dining tables, ready for our weekly Sit Rep, i.e. Situation Report (all words are summarily guillotined in the Antarctic).

For weeks I had been lobbying the Operations Manager to get out into the field. The taxpayer had sent me to Antarctica to write about it, but so far I'd been shuffled off the roster by more important people, like government ministers and journalists from television networks with 6.7 million viewers.

So I was pleased to see the Ops Man write my name on the white board, under 'Ice Blu' ('blue' being clearly too long to stomach). This was the name for a fuel depot and logistics base in the Ellsworth mountains, a thousand kilometres south of our base on the Antarctic peninsula. Beside my name was Drew's.

Drew was one of the five pilots based at BAS's operations hub on the Antarctic peninsula. I'd met him already on several occasions. He was friendly, accommodating. The night after Sit Rep, Drew took me to the Air Unit office and turned on the computer. Images slid across the screen. Ice, sky, mountains, ice, sky, mountains, as if they were all the same substance.

'I want you to see where we're going, in case the vis isn't so good,' he said. ('Vis' meant visuals.)

'But if the vis is no good we won't go, will we?' I asked.

'No, we'll still go in. I can fly it on instruments.'

Drew's voice had the resolute, unarguable timbre of a pilot. I didn't know how old he was but I guessed fifty-six, fifty-seven. His voice was not worn or shredded at the edges, as the voices of people of his age often were. His was still smooth with something that might have been hope.

He flew into Ice Blu only a few times a season, he said. He was usually sent to the Ronne Ice Shelf to the east of the peninsula.

'How does it differ, flying into these two places?' I asked.

Drew told me how he would fix his approach from the ice shelf, flying low over the broken playground of ex-continental ice: frazil ice, lacy ice, ice flowers – whole fields of them, smashed into hexagonal pieces of muslin. He loved the approach onto ice shelves from the sea because it challenged him as a pilot; the ice cliff was so high, and its edge weathered.

'You have to get her on the deck just right. It's wind-scoured, so there's always sastrugi.' Sastrugi was waves of frozen snow. It meant a bumpy landing. 'If you miss the approach well…' With his hand he made an abstract gesture.

He told me about places he flew on the continent: the Ronne, Filchner and Brunt Ice Shelves, the South Pole, Patriot Hills, the Ellsworth Mountains – which we would see, luck willing, the following day. Such thrilling, triumphant names. Yet it was *Bluefields* that would not leave my mind. It was just another fuel depot, an anti-place with nothing to locate it apart from a few poles with black banners sticking out of the ice. There were no fields, Drew said, and no blue, except of course the sky – on a good day.

'Why is it called Bluefields?'

He shrugged. 'Who knows? Names in the Antarctic are strange.'

Antarctica is the only continent without a history of human habitation. The names of places have been grafted on. Taken together they have a utopian thrill underscored by dread: Deception Island, Pioneer Heights, Dismal Basin, Turbulence Bluff, Cape Wild, Quest Cliffs, Minerva Glacier.

He shut down the computer. 'I have to go talk to the Met Man [i.e. the Met Office forecaster] and get the latest.' He called over his shoulder, 'We've got a long day tomorrow. Be sure you get a good sleep.'

My eyes were glued open all night. The following day I would fly, alone apart from a man twenty years my senior who I had met only the previous week, into an immensity. If something went wrong and we had to ditch the plane, there would be a search, but possibly no rescue. Sometimes it wasn't possible to save people in Antarctica: that had been drilled into me since I began working with the British Antarctic Survey some four months previously. We might have a scenic over-flight of the continent and see things few humans will ever witness in their lives, or we might die together, probably of hypothermia or dehydration or both. I didn't get much sleep because my brain went straight into fiction mode, drafting a Jack London-worthy survival story in which I expired first, probably from boredom, and Drew hunched over a pot warmed by the last of his methylated spirits, boiling up my hamstring.

We took off at 6am and were snagged on an updraft, rising a thousand feet in seconds, a reverse-vertigo common in the Twin Otter, a tough, rustic bush plane. The cockpit had only a small electric space heater so we were bundled in Andean ear-flap woolly hats and BAS-issue RAB mountaineering puffer jackets that made us look like Michelin Men.

We flew for ten hours, flying, refuelling, unloading, taking off again, finally landing at Ice Blu at the edge of the Ellsworth mountains. Because I had arrived in Antarctica by ship, this was my first sight of the interior ice sheet from above. Every so often Drew would gesticulate to put my hands on the throttle. I would keep the Otter steady while he took photographs and recorded our position. The plane shuddered in my hands. When Drew took control back the plane felt happier, like a horse that knows it no longer has an amateur on its back.

I kept notebooks stuffed in all possible pockets in Antarctica, the way some people hoard chewing gum or snacks. I managed to extract one of them and scribble, my hands muffled by gloves. I kept writing, or trying to, about what I witnessed. But language refused to organise itself around the land, or whatever it was we were flying over, because it looked like no land I had ever seen before. Here is a fragment from that day:

The light – barking, infinite. The sun beats down on the blue-green-white glare of the snow, shocking against it, bouncing back, not a sun but cosmic rays. The air shimmers with the certainty of the encounter. The Antarctic goes on and on, billowing into the horizon, a giant white jellyfish.

After a while I stuffed the notebook back in my Michelin Man suit and sank into a meditative state induced by the searing blankness underneath us. I took photographs out the scarred windows of the Twin Otter. They tell their own story: the black tors of nunataks, like chess pieces dropped in a snowfield. Rolling waves of ivory, wind-smoothed, a frozen ocean. Blue scallops of moulins. Crevasse fractures spidering across a river-less land. Pleats of ice. Even in its two-kilometre-deep sarcophagus of ice, the land broadcast upwards a buoyancy, as if it could rise at any moment and nudge us out of the air in our red flint of a plane.

I will never be able to write about this place. The thought installed itself in me, in the cold cockpit of a Twin Otter, ten thousand feet above the Antarctic's Europe-sized ice sheet. It was a sobering realisation, considering I had been sent here to do just that.

Drew's voice cut into my radio headset – the sound of the wind and drone of the engines made talking, even at full shouting volume, impossible. 'Enjoying the view?'

I turned to face him. He must have seen my expression before because he smiled, nodded slightly, and went back to looking out of the window.

Magic!

Two silhouettes sketched themselves into the horizon: Pete the radio operator, Mike the comms manager. We ski across a snowfield as flat and percussive as the bonnet of a car.

We have lived in the same place, Rothera base, for months, but I wouldn't say Pete, Mike and I know each other. I don't know if they are straight or gay, how many brothers or sisters they have, if their backstroke is good or whether they have ever been ice-skating, where they were born of if they were presented with rabbit if they would eat it (in any case this isn't going to happen in Antarctica). Yet we are intimates. We all see each other every day at breakfast, flecks of unnoticed toothpaste wedged in the corners of our mouths, we can feel each other's sour Tuesday afternoon moods coming like weather, we share sudden bursts of uncontainable energy after eating a chocolate bar and we all run out into the new snow to watch passing Southern right whales at the end of the runway, marvelling together at their slick, lapidary backs.

It is March, autumn in the southern hemisphere, and we are all in the denouement phase of the same old story, the one about time winding down. We are in the Dog Days of upside-down summer, the final performances of the run of an acclaimed play, we are filming the pick-ups in a movie about to wrap.

After four months in Antarctica, it occurs to me I might actually leave. Antarctica is a jail but an intermittently thrilling one. Soon this place I sometimes resent and want to escape will become inaccessible to me, forever. The thought brews an abstract hunger I have never felt before, for a place I actually am in but from which I have already been severed, except I haven't, because we have to live out these final weeks on base with their promise of finality.

At breakfast we concoct expeditions to pass the time. When Pete and Mike issue a generous invitation – I am not a hardy mountaineer and I might slow them down – I snap at it.

Our mission is to climb Stork, a mountain ten kilometres from base. It has a commanding view high above Adelaide Island and for months now our only views have been of immediate surrounds of the base: Ryder Bay and the mountain-islands of the section of the Antarctic peninsula called Graham Land.

Winter is on our doorstep so we have set out early, to make sure we have plenty of rescue light. We carry coils of heavy rope slung over our shoulders and our waists are swaddled with a heavy belt that would be the envy of any sheriff in a Western. Carabiners clink in the silence as we walk. Our heads are encased in shiny red helmets. When we catch glimpses of each other, distorted in the mirror of our wrapa-round sunglasses, we look like giant Gregor Samsa insects.

As the sun gains strength we scramble up the mountain. It is hard going. I've kept myself fit on base by running multiple laps every day on our nine-hundred-metre-long runway, but my thighs are not used to jack-knifing on frozen rock while carrying twenty kilograms. We have crampons but still my feet slip every three or four steps. My shins take the brunt of ice and rock-fall, knocking painfully against the frozen mountain.

After an hour or more – I lose track of time – we reach the top. We sit, digging ourselves hollows of snow as windbreaks. It is not a cold day, only minus eight. The sun glares but emits no heat. Underneath our layers we are sweating from the effort of the climb, and if we stay still for too long the sweat will cool, and we will catch a chill. We break open bars of twenty-five-year-old chocolate (all expedition food in the Antarctic is decades past its end-date but in the dryness of the conti-nent does not degrade) and chew. 'Not bad,' Pete says, not of the view but the chocolate bar. The bar has lost the bite of its cocoa, but we eat it nonetheless. We will need its energy to get us back to base.

Below, snowfields terminate in the wrinkled folds of hummocks – pleated runs of ice and snow at the hem of mountains. It looks like a bowl of meringue. There is no sound at all, apart from the whisper of snow sent skating across itself by the wind and the low hum of blood zinging through our veins. Before coming to Antarctica, old-timers I'd met at the BAS pre-season conference had told me about the silence. 'You won't hear anything but yourself.' I realised I'd never before listened to the sound of my body. It was like a low-pitched string on a violin, a B or G flat, one of those resinous notes that sound like rain.

'How did you find the climb?' Mike asks. He is the quintessential comms man; bespectacled, precise, with a PhD in Physics (everyone in Antarctica is overqualified).

'Hard.'

'You'll get the climbing bug if you're not careful,' Pete says. 'You'll find yourself hanging off bouldering walls in Walthamstow or wherever you live.'

Stoke Newington, I think, but do not say, in part because the very idea of my neighbourhood is absurd, surrounded by this cathedral of snowy mountains and the knife in the air. And I always thought climbers were certifiably crazy, risking their lives under unstable seracs and bivouacking on ledges, braving vertigo falls and crevasses and other forms of certain death. But now I realise that mountain climbing is a reverse Orphic quest. The mountains beguile susceptible spirits to climb them, and up they go, vertigo pilgrims, into the threshold between this world and the next.

Pete's gaze rakes the sky for signs of change. The cloud that has taken hold in the sky was thin but enough to buffer the sun. The sky has lost its eggshell patina of blue. We begin to feel cold. We emerge from the snow forts, stand, stretch aching muscles, and take in the view one last time, before heading down.

Mountains ripple all around us, throbbing magnetically in the silence. To the west is the edge of Adelaide Island. The distant open ocean glints like chrome, a sign of pack ice, or a mirage. I think: this is what time looks like. Slow, implacable, infinite. A duration of rock and silt and snow. Not for the first time, I try to absorb the bare fact of Antarctica. It is the only continent that has resisted, long-term, the alien invasion of us. The victory in the land is audible. *See, we don't need you at all.*

For a moment, fleeting and cold, I am absorbed in time, no longer its chronicler but its artifact. Dinosaurs lived here in three-month-long darkness, nudging between giant ferns and clubmoss trees, subsisting through the months-long night, right where we stand, eating our newly-ancient chocolate, on mountains yet to be erected.

'I'm feeling strange about leaving. I want to go, but it feels like I'm signing up to death – or something like death. But that to stay here would also be to die. So it's death all round.' I try to laugh but it chokes in my throat.

Steve nods. 'Antarctica is a once-in-a-lifetime chance, for most people. Leaving it feels like mourning.'

Steve the base commander and I are drinking coffee in the clotted twilight of what was so recently morning. We are losing an hour of light every day. Because of the tilt of the planet, the light advances and retreats with the seasons faster than anywhere else on earth, including the Arctic. This rapid repeal is something the human body will never experience anywhere else. My body knows something is wrong and has activated a low-level panic.

'But you don't feel that?' I ask.

'No, I know I'll be back. I've done four winters here. When I leave I say goodbye to home, knowing I'll return. Every time I go Antarctica says, see you next time, you're welcome here.'

'How is that?'

The laugh-lines around his eyes crinkle. 'I know, it sounds mad, doesn't it? But I've heard lots of people say the same – the scientists, the mechs [mechanics], even the pilots, who are professional nomads. They all say it feels like home. That's another thing that drags people back here,' Steve says. 'They need to get rid of that sadness. The only antidote is to return.'

Steve looks out the window, toward the runway. A red shadow crosses the view. The last remaining Twin Otters on base are coaxed out of the hangar by the base tractor. This is our signal to don our puffer jackets and mukluks and head out to the runway.

Today the Twin Otters will leave on their long hopscotch trip to Calgary, via Stanley, Florianópolis, Belém, Caracas, Guatemala City, and somewhere in Texas. I couldn't remember the last stop – the pilots had listed their itinerary in the bar the night before; Wyoming, maybe.

The departures of the Otters draw every person on base (apart from Mike, the comms man, who must stay in the aircraft control tower to despatch them) to the runway's edge. It is a mild day for late March, around minus two degrees. The plough has been out on the runway overnight, clearing freshly fallen snow. After the planes leave the snow will be left to accumulate and I will have to clear my own path to continue running.

The day is overcast but the Antarctic light is still strong enough we clamp on our sunglasses and watch as the two Otters, one after the other, line up on the southern end of the runway. There they shudder, the pilots revving the propellers for their take-off run. Within seconds they are airborne, rising resolutely in a stiff headwind. We watch as they disappear into the sky.

We file away, back to our offices and workshops. It is hard not to feel abandoned. Now, everyone on base will dive into a frenzy of activity to keep ourselves busy and distracted from our predicament. With the planes gone, the only way off base is by ship. In two weeks' time the *RRS Ernest Shackleton* will come to collect us, battling through autumn seas from the Falklands.

I have a dilemma; everyone expects me to be at my desk all day, writing. But writing is the last thing I want to do. Instead I spend hours eavesdropping on the field assistants, the mountaineers who support the scientists on off-base expeditions. These men (they are all men) talk about grips, crampons, carabiners, they punctuate sentences not with full stops or explanation marks but words that serve the same purpose: *gnarly! Awesome! Magic!* Or I help the base general assistants cart tins of mushrooms or tackle learning to drive a JCB (not easy). Anything other than write.

The books lined up on my single shelf look convincing enough for an Antarctic writer-in-residence: *South*; *The Worst Journey in the World*; *Robert Falcon Scott*. I have carted these classic tomes all the way from Britain. In London, researching the Antarctic in the months before I joined the expedition, I devoured these stories of hardtack and hoosh, of killer whales that nearly did kill them, of destroyed ships and frostbite and being lost to the world in an age before even telegrams existed. But here, if I try to read them, I feel a sharp pain at the bottom of my lungs. I keep trying but after a couple of pages I have to put them down.

My residency with the British Antarctic Survey is a very official appointment. I've had to get permission from the Foreign Office to be here, undergo a battery of medical tests and fill out a psychological questionnaire. I've signed waivers and submitted proposals about what I intend to produce as a result. But I have never written to order before, I have never self-declared the subject and purpose of a book before I have researched it. For me, writing fiction has always been like setting out into a dark sea in a small boat, destination unknown. To have a brief and a subject ought to feel safer. I should have an anchor, a legitimising purpose. Yet I sense a trap.

At Rothera I am paralysed and mesmerised equally. Not only by the landscape, but by the combination of danger and mundanity our lives have taken on. We do the same things, over and over again, against a backdrop of restriction and peril. Putting on your boots, taking your boots off, hanging your boiler suit on the same hook each day, two slices of toast with Marmite, disappointing coffee, watching the fur seals slump and thump at the end of the runway, the scar of a skua as it tears through the sky, moving your tag from IN THE PIT ROOM to BRANSFIELD HOUSE, the turquoise nights.

We have long since accustomed ourselves to the atmosphere of impossibility. We can't go home. We can't eat cheese that does not crumble (an effect of freezing). We can't have an espresso or walk in a park. To even take a stroll around base presents multiple hazards. We might misjudge the edge of sea ice, fall through and freeze our foot off. Anything can happen here, in this chamber of waiting and silence.

I have another problem, one which has dogged me since that trip in the Twin Otter with Drew. I have been failing to be able to describe what I see around me and what we do, even the most basic things: the sky, the sea, the mountains. My failure to render the Antarctic is not because I lack the vocabulary. I have the maps and the *Mariner's Encyclopaedia of Ice Terms*, and thousands of words of notes and hours of taped interviews with glaciologists and atmospheric chemists. The problem is that being here doesn't seem to help me interpret how it feels to be here. I have realised too late that the only meaningful thing to write about the Antarctic is that it exists.

To divert attention from the not-writing I learn to play the role that will become my specialism in my years of attaching myself to science expeditions, that of witness-sidekick-dunce. I accompany marine biologists in Rigid Inflatable Boats as we speed around Pourquoi Pas Island to collect sea spiders. I tag along with oceanographers to sample the water column, then fly out into the ice sheet with glaciologists to listen to ice streams draining from the frayed edges of the continent. I stage late-night conversations with ice core drillers, atmospheric chemists, physicists, terrestrial biologists.

I learn that marine life is struggling as the sea temperature inches incrementally upward. Sea sponges, sea cucumbers, sea spiders all have a thin capacity to adapt to warming. Like coral reefs, a heating and increasingly anoxic ocean will finish them off. Grass is growing on the tip of the peninsula, the terrestrial biologists tell me, for the first time since the Precambrian era. The bubbles and isotopes in ice cores are witnesses to what we already know: we are conducting a carbon experiment, unprecedented since fifty-six million years ago, when the earth abruptly went on a carbon-emitting spree for reasons known only to itself, releasing more CO_2 than our global fossil fuel reserves over five thousand years. The thallium, sulfur and nitrogen in the Antarctic's rock are trying to speak to us, to tell us of how the ocean lost much of its oxygen in this interval, becoming a dead zone, an anoxic soup. The scientists are the translators of this language, but now they have a dual purpose, they are in fact interpreting two languages: that of the past and that of the future, of what the earth will look like, when we are done with it.

The Last Glacial Maximum

In the final days of March the weather clears to reveal a steel sunset. Shoals of ice have been shoved into the bay on the tide, where they are broken up by the local gyre like giant sheets of scrap metal.

A single fur seal, its back arched as if in supplication, stares up into the clouds. The late autumn horizon is rimmed with sallow light. An iceberg lies grounded in the bay, a broken buttress, trapped in the grandeur of its disintegration. The tobacco sunset gleams through the giant hole in its middle. A strange moon installs itself in the sky, red-gold, an autumn harvest moon made ruddy and weird by the latitude.

I get up from my desk and walk out the door. I am in shirtsleeves but at least I have remembered to put on my rigger boots. Patches of ice cling to the veranda. Across the bay is Léonie, a stark triangle-shaped mountain, small by Antarctic standards; its basalt flank is a key landmark for pilots when they land.

Icy stars are suddenly visible, as if switched on. The Southern Cross tips into the edge of the sky, revealing the smudge of the Magellanic Cloud. There are no familiar geometries, only a jumble of pulsating green stars against a blackness of unfamiliar density, as if we have been transported to another planet.

Cold ruffles the edges of my body. I wrap my arms around my chest.

Rory, a field assistant from Northern Ireland, comes out of the mess hall. He taps me on the shoulder. 'Aren't you a bit chilly?'

'I don't know.'

'It's minus ten.' He frowns. 'Come on, I'll show you some pictures of my trip to Berkner Island.'

Rory's photographs are of the ice shelf base, Halley. It lies 1,800 kilometres, nine hours' flying in a Twin Otter, northeast of Rothera, perched on the Brunt Ice Shelf. Halley is much more inaccessible than Rothera. At Halley a ship – either the *JCR* or the *Shackleton* – does Relief in late December, when the changeover of personnel takes place. Other than flying visits from Twin Otters, that is the end of its interaction with the outside world. If Rothera looks like a military base, Halley is fully extra-terrestrial. A rust-coloured industrial platform hiked twenty metres above the ice surface, it could easily serve as the set of a Mars colony blockbuster.

We are living on a new planet, I think. Or, we are astronauts, orbiting Earth while still on it. This dislocation has installed itself in me, just by living in the Antarctic. I don't know it then, but it will never leave me, the sense that I once left planet Earth, and now I know what it really is: a notional realm. The Antarctic has the capacity to collapse the physical and the metaphysical. Thinking about it in that drab laboratory with Rory, looking at his interstellar pictures, I feel that lateral vertigo I will sometimes feel – on ships in the middle of the ocean, or walking across the Namib desert – for the first time.

Afterward I go to sit in the base library. Its windows are plastered with snow. Outside, the mountains and islands that surround Rothera like sentinels imported from Norse mythology are tipping into a purple darkness. It is 4.30pm. My northern hemisphere body fights back, knowing that it is soon to be April, and spring should be on the way. My body is hungry for green, flowers, the platinum light of England. It rebels against being pitched into winter in the tipped-over seasons of the southern hemisphere.

The stirrings of a new, unfamiliar kind of anxiety take hold. In coming to Antarctica and marooning myself in these soupy last days on base before winter, I have cut myself off from the ordinary world. That much I understand. This time, my unease is not generated by me, or not only. It is a collusion between myself and the landscape. My body seems to know that it should not be here, and at the same time I receive the same message from the land: *go away*.

Léonie is lit in Silverprint light. Out there, nothing for as far as I can see, and ten thousand kilometres further after my gaze runs out, is sleeping tonight. No land-dwelling creature – not even emperor penguins, which in any case are stationed on the other side of the continent – is alive to see this sharpened light, the note of finality in it, so extreme it almost looks like an expression of infinity. As darkness takes hold of the continent Antarctica is solemn, resolute. It is hard to believe anything humans might do will disrupt its throbbing propriety, its confidence.

Now I can only sleep a couple of hours a night. Nothing matters except sleep, and yet it refuses to come. My head zings with anxiety. Every night I linger in a malign hinterland of sleep, watching the hours slouch by: 11.30, 1am, 4am, 5am. Time has been subsumed by a dark glistening substance, like liquorice. I am convinced that this element which has replaced time is the same as the black blood running through my veins.

What happens to me in the final chapter of my Antarctic adventure reminds me how little I know about myself. I feel out of control, subject to forces over which I have no dominion and little understanding. These forces surge through me, from within me, but they are outside of me. They are everywhere but dispersed. I feel poised on the edge of a collapse, or a cliff and all past and future, all reason and cause and effect, has taken on the appearance of a dark mystery. In such episodes, you have only three allies: your will, time and friends. These are the only things that will save you.

There is no moment in which I think: the Antarctic is speaking to me. It is a gradual impression. I don't think the continent is speaking exactly, and not to me, either – I am not that grandiose. But something is apparent that was not before. A presence. Or no, that's not the right word.

What I feel is an incline in the energies that I have taken for granted all my life and which are generated by a confluence of seemingly unrelated things: time, the air, possibility, the land we walk on, birds, the notional future, the sense that I am in a realm that might not exactly welcome my existence but which has decided to tolerate it. There is a thickening. It feels like the presence of a will, and it is coming from the land, as in from the ground we stand on, even if the ground is buried under two kilometres of ice.

In those countdown days to our departure, I go for solitary walks around The Point. This is a circuit we tread around the small peninsula. Heading north, base is behind us; as the route curves toward Ryder Bay, it disappears. It is the only place we can go to not see the comms tower, the hangar, the outbuildings, all buttressed by loafy snow. Briefly, we can forget base exists.

With base vanished, I can concentrate on listening for the source of this sound. Icebergs the size of small ships stall in Ryder Bay, the sun striated by holes in their hulls. Pourquoi Pas Island beams a disapproving frequency in my direction. It is this I pick up here, mostly: an almost parental admonishment, emanating from the land. The sun shifts, somewhere within the slabs of Antarctic cloud. Water laps against the serrated ice that lines the peninsula. Otherwise, silence – no birds, no wind. I strain to pick up the frequency I felt moving through, but it has gone.

Always on these walks, a strange thing happens – when I put base behind me, I am relieved. Finally I am alone in the wilderness. But after twenty minutes I begin to want to see its khaki-coloured panels again, to witness the human figures waddling between the laboratory and the mess hall, the wires that zing our satellite signals to the outside world. By the time I round the corner at the southernmost point and the satellite dome comes into view I am relieved all over again.

Back in the mess hall that afternoon, the Ops Man gives us bad news. The *Shackleton* has engine problems and must remain in the Falklands while spare parts are flown from the UK via Uruguay. *How much longer?* we chorus at the base commander in our weekly Sit Rep. He shrugs. 'This is Antarctica.'

There is a dull ache inside me, somewhere around the bottom of my lungs. At first I think it is physical, some effect of the dryness of the air, the cold. In the science lab next to mine a calendar is tacked on the wall: *Trees of Wales*. The calendar girl for March is a mulberry tree. The band of pain tightens around my lungs as I flip through the trees that star for each month: linden, oak, willow. For the first time in my life I've been living in a place without trees and it's as if someone has died. Is that why the land here is so silent? Without the low buzz of trees' communication networks, their stringy rhizomes like underground mobile phone towers just underneath the threshold of human hearing, the very air feels empty. There are no above-ground tree-sounds either: the wash of their leaves, the creak and sway of branches bending in the wind. I think of the trees I grew up amongst, Scots pine and fir, mostly, so thick they made a mesh that sealed us into ourselves, the highways of my childhood that ran straight-shot through them like arteries in a body, the trees murmuring, writing an oral epic, winding all around the world.

In my final week on base a panic coagulates. There is a pain at the bottom of my brain that throbs menacingly day and night. Nothing – no antidepressant meted out by the base doctor, who has seen this kind of unravelling before, no sub-set of Valium – can quell it. It sounds like someone welding metal. *Screech, scratch.*

At night, on the rare occasions when I can sleep, I fall into baleful dreams. A Viking-type face (horns, braids) scrutinises me from the other side of my eyelids. A woman with the eyes of an owl peers at me, even when my eyes are open. She sits in the corner of my office, casting looks at the purple glacier above base. Owls and Vikings must be able to handle the end-of-the-world lostness of Antarctica. Maybe they have been sent to help me survive whatever is happening.

The distress I feel ripples through every cell of my body. I start to walk in circles around my office, then to pace it on the diagonal. The generator mechanic in the next office trying to talk to his family on Skype loses patience with the squeaking floorboards and bangs on the plasterboard. I sit down and put my head between my knees.

The days creak by. We are bolstered by routine: breakfast, morning smoko, the daily Team Timed Crossword, lunch, afternoon smoko, dinner, blockbuster films in the bar. But there is an air of exhaustion that settles in quickly. In the bar people stare silently into their cans of Guinness.

The wind groans. Stepping outside is like entering a blasted open-air cathedral, the organ wind grinding out some nameless Requiem. *Chances are you'll never be here again.* Steve's words rotate inside me. I've never been good with nevers. As soon as anyone suggests that something is impossible, or never to be repeated or understood, I set about proving them wrong. It's contrariness, but also a recoil at the diktat terrain of neverland: how do you know? Never is a very long time. People love to dish out nevers, they are like sweets lightly dusted with poison fed to a tyrant. Everyone loves to imagine they are in control of time, as if they already inhabit the future.

The Far Field

The ship was scheduled to arrive at 10am. At 8.30 an announcement went around on the bing-bong, as the internal PA system was called. 'Shackleton visible, I repeat Shackleton visible.'

We struggled into our orange insulated boiler suits and waddled up to the viewpoint over base called The Cross. A marine twilight stood in for what would normally be called morning.

At first it was only a dot, almost indistinguishable from the sky, on the other side of a large girded iceberg that had grounded itself in the bay.

The steel-blue air around us tightened. I heard a faint signal. I realised it was not a sound so much as a lack of silence. With the ship nearly upon us we were about to lose the sound of our solitude.

I shuffled through the snow to my office for the last time, to box up my bookshelf. Sitting among my Antarctic heroic volumes were books that had nothing to do with the place, but which for me were literary talismans: *Eros the Bittersweet*, by Anne Carson and *The Far Field*, by the American poet Theodore Roethke.

As I packed, I thought of Roethke's life. He was a big burly bear of a man, always clad in fraying jumpers. In photographs he looks like General von Clausewitz. A *New York Times* article on Roethke's work published in 1972, eight years after his untimely death, is titled 'As if haunted by a raging dark angel'. For much of his life Roethke ricocheted between euphoria and depression. The intensity of his vision and feeling made ordinary life sometimes impossible for him. He struggled to keep teaching positions at university, drank heavily, and had spells in mental institutions.

'In a dark time the eye begins to see,' is the opening of one of his best-known poems which takes its title from the first four words of the line. Roethke described himself as an 'eyeless starer'. He found a wintry immanence in the soul-night of depression. Blind angels, rays of light, humming, lordly Rothko-like horizons – just as in the painter's vision, Roethke's work is replete with Silverprint limens. His rapturous rendering of the streams, stones, ponds, forests, islands and crags of his native Michigan serves as a reminder that the task involved in writing about the more-than-human world is simple, in a way: to be present, to pay attention, to see the life in seemingly lifeless things. To see the shadow of the beyond-realm, the *there* space.

Most of us never get *there*, as writers. We lack the vision and the courage. But beyond this mundane failure is the greater disaster that language is not enough. The search for language is always a quest for that which is just beyond the grasp of one's expression, but not perception. We perceive many things for which we have no words, or at least the words are an inadequate match for the feeling. Mostly our search ends in vanishings or near-sightings, like the tail of an animal glimpsed as it melts around the corner. We need a new idiom, one that works across all speech, a specific, stern, vaulting tongue, that recognises we are only itinerants in this realm, a language that will finally reveal our debt to the world.

We departing summerers prepared to board the *Shackleton*, leaving the twenty-one overwinterers behind. We filed onboard and went straight to the cafeteria to stare dumbly at wonders called avocadoes and lettuce.

The next three days were frantic with unloading and loading of the ship: we carted boxes of tinned mushrooms, crates of beer, enough tinned asparagus and frozen chicken, to last the seven months before the winterers would see anyone from the outside world again.

When we were finally loaded and ready to leave, the winterers lined up on the wharf in their red BAS-issue jackets and fired emergency flares into the air. Their orange burn made it look as though base had been engulfed in a hungry, anoxic fire. The *Shackleton* sidled away from the wharf, almost reluctantly, its bow thrusters churning. The ship pirouetted on its haunches. The base, its matchstick outbuildings, the snow-path we ploughed ourselves from Bransfield House to the wharf, the satellite dish, its lunar orb, receded. From afar, base looked like a giant freezer in need of defrosting.

I thought of the never-filter Steve and I had talked about. My eyes willed the Antarctic to etch itself upon them an image – Turner's greasy pastels, a drawing of Dürer's uncanny precision – something that had the heavy certainty of all time about it. But this, too, Antarctica resisted. All I saw was the back-of-the-freezer detritus, and the sky's inky, unconvincing attempt at morning.

If never-vision had been available to me this is what I might have seen: all the years the continent has been expanding and contracting, a jellyfish moored at the bottom of the planet, swimming through the long ocean of time, turning its nervous system on, off, on, off, dormant in the three-month-long exile from the sun. The Antarctic really is like no other place on earth. Its completeness, its hermeticism and grandeur, never mind its moral authority, is unique. It carries the gravitas of space, it speaks in the voice of the planet before us, when it was secure in its aloneness. The Antarctic is a frozen nebulae crashed on a wet planet, its ice burning the mouths of every one of us like fiery pastilles. It is the scene of a reverse fire, the most convincing vision of eternity I have ever seen. It is ice and days and ice and days, the far field of a time without us.

On that winter morning in 2006 it was not yet clear how much of the Antarctic we were going to lose. We did not know that (according to NASA) forty-nine thousand gigatonnes, or one billion tonnes, of polar ice has melted since 1901. That, even if we were able to meet the vanishing target of 1.5 degrees warming, as the 2015 UN Framework Convention on Climate Change conference in Paris stipulated, we are probably going to experience sea level rises of at least thirty centimetres by 2050. We did not know that we are now 'locked in' to this scenario, no matter what happens. If you take into account thermal expansion and the possibility of methane release, the oceans will rise perhaps two metres over the next five hundred years, or even more. And if the entire Antarctic ice sheet melts? A minimum of sixty metres sea level rise. Enough to engulf most coastal communities and cities around the world, as well as submerging all low-lying islands. Enough to mean that most of us are living in future Atlantises.

Living in Antarctica had been like sharing a hotel room with God. Antarctica is here for what anthropologists call the *longue durée,* and I am a guest for barely one night. Ordinary human feeling – emotional experience and attachment – was not up to the Antarctic. There, all emotion becomes feathery and self-regarding. That morning in April 2006 the only feeling I could muster on leaving Antarctica, probably forever, was a hollow kind of possession. I imagine it might not feel so different to leave life itself.

We were the last ship heading out of the continent that season. In the mauve gloom of the morning we set course for the tip of Adelaide Island, heading north, back to what everyone in Antarctica called the real world.

PART VIII

THE RAINY SEASON

October 7ᵗʰ, 2020. London, UK. 14°C

What I did today: four Zoom calls, or was it five, followed by a flurry of increasingly frantic emails trying to get airfares for pandemic-cancelled flights refunded.

We've been living under house arrest for nearly seven months now. My life has been laid bare, like a curtain on a stage, lifted, to show a Spartan set. I never wanted to be married or have children. At least I knew what I didn't want. I knew I was lucky to have been born into a rare moment in history when a woman did not have to either depend upon a man for her economic survival or breed. This seemed to me such raw good fortune I had to act on it.

As for writing, there was never any decision. This is how it happened – I think.

I was eighteen years old and working in a restaurant – one of the top restaurants in Toronto of its day – to fund my university tuition fees. We regularly served a glamorous publisher. One day a job came up at her publishing house and I applied. Soon I found myself collecting Kazuo Ishiguro at the airport, accompanying John Irving to hockey games and even more improbably doing the production edit on an anthology of Canadian fiction edited by Michael Ondaatje.

Sitting in as the writers we published gave interviews to newspapers and radio stations, listening to their literary festival appearances or just talking over lunch or dinner – the late 1980s and early 1990s were a brief interregnum in the publishing industry when everything was not about money and the whole editorial and publicity team would be invited to these – I had a live demonstration of what being a writer meant. Writers were chroniclers of the present before it evaporated, certainly. They were bellwethers of the future. Fiction in particular was a charged, symbolic realm, a zone of the realer-than-real, a more meaningful compendium of truths than any other form.

Nothing in my family background or life to that point suggested I could be a writer. But I was inspired by the writers I was fortunate enough to meet, all of them at the zenith of their powers and careers. A *why not?* energy began to circulate within me. When I wrote poems and stories they were immediately published in Canadian literary journals, of which there were helpfully many, supported by an array of provincial and federal grants. It was a different chapter in history, when you could subsist on a diet of accident and luck, as opposed to now, when life feels like being an insect stuck in flypaper.

In the early 1990s I moved to London and started writing travel guidebooks to countries in Latin America, mainly Central America. Then, you could almost live on advances. As long as you carved out an audience and received good reviews, you had a chance. Then, the act of recording and interpreting the natural world appeared both honourable and necessary, if not yet an emergency, a mission impossible. Ours was the business of interpreting the world, not writing its requiem.

In those years climate change was still a largely scientific preoccupation. The seed of a new way of thinking about the world was being planted. Environmentalism and conservation were the parents of climate change. They made it possible for the large-scale destruction to come to be understood. My years in Central America in the mid-nineties was my first total immersion in an unfamiliar landscape in order to write about it. I had to learn it by rote – all the birds, flora, animals, its history, its flaws and its hesitancies.

As the pandemic days shuffle by I try to remember those years and find I can only conjure a few images. I had thought memories kept their structure over time but lost the richness of their details, but I find it is the other way around – I cannot remember cause and effect, or where or why something happened, but specific moments remain. 'Memory is not an instrument for exploring the past,' Walter Benjamin wrote in his essay 'Excavation and Memory'. Rather he likened memory to the place where the past is buried. What we commonly think of as remembering is a process of excavating the thing itself, he suggested, akin to remembering our memories, more than 'the past'. We do this through images, Benjamin argued. Our memories of memory appear as flashes, fragments. If I close my eyes, what do I see of those years I spent commuting up and down the slim isthmus of Mesoamerica?

Sun falling syrupy from banana leaves. Gigantic spiders on my pillow. Garrisons of a translucent, almost plastic, green braided by lianas and studded with pink parrots. A carnal equipoise between rainforest, mountain and ocean, until they were all the same body. The underwater UFOs of green and olive ridley turtles, delicately paddling their bulk through the sea underneath me as I swam, turquoise with promise.

Monteverde, 10.2989°N, 84.7682°W

Cloudforest

A slight figure emerged out of the mist and a pale dawn. From afar it was hard to tell if it was a man or a woman. The figure wore a baseball cap, a long-sleeved shirt and those American-made walking trousers made of a newfangled material at the time – Gore-Tex, probably – stuffed in Wellington boots at least one size too large, all underneath a see-through rain poncho.

'Buen día, Jean.' The voice had a light timbre, like birds. She pronounced my name *Heen*, as is logical in Spanish in which the hard 'G' sound is softened. We shook hands. Hers was firm, resolute – not entirely a woman's handshake. In the half-light and because they were cast in shadow under her hat, I couldn't see her eyes. She looked young, with a child's smooth oval face.

She pulled down her hood. Marisol came into focus. A small woman, no taller than me, wearing John Major glasses, just as I did, and as was the fashion (when and why did glasses shrink?).

I'd made contact with Marisol via the information centre in Monteverde, one of the largest cloudforest reserves in the Americas. The centre also functioned as matchmaker between tourists and guides. I'd requested a specialised person who could give a scientific account of the cloudforest, but to a non-scientist: me. *'Quisiera una mujer?'* The man who attended to me at the centre asked. Would you like to be guided by a woman?

227

Until then, all the guides I'd met had been men – genial and respectful as all of them had been, I'd tired of the wary suspicion with which they met me, a twenty-four-year-old woman with no husband and no children, which in the culture of the time and place put me on the same plane as a forgotten mythological creature.

I would discover Marisol was only four years older than me and, at twenty-eight, already had two children as well as a PhD in tropical biology from the University of Costa Rica. I explained to her that I had no degree in tropical biology or children and apologised for both as well as revealing I had approximately six months to write an authoritative account of the whole country, including detailed sections on all the Life Zones found in Costa Rica.

She gave me a level look. If you are attentive you can sometimes see people's entire thought pattern in their gaze, as visible as a silhouette in a window. I saw her cycle through shock, dismay, disapproval, to arrive at pity.

'Okay,' she said, finally, a determined note ringing in her voice. We were to start with two walks, one in the morning and another in the afternoon. We would continue until I could recount to her the basic biology of the cloudforest biome.

'I don't have to be an expert,' I tried to reassure her – or me. It was a phrase my editor at Rough Guides had offered on more than one occasion. *You just have to know enough to write about it.* Neither of us seemed to realise that in the case of tropical biology this might mean knowing rather a lot.

We started down the road to the entrance of the National Park. The Monteverde Cloud Forest Biological Reserve is the most visited cloudforest in all the Americas. It is twenty-six thousand acres in size, with thirteen kilometres of trails which need to be continuously rebuilt and maintained against rain, mudslides and constant use. Dairy farms and smallholdings nudged up against a solid garrison of cobalt trees. Mist swirled around our feet – I'd never seen that, at ground level, other than the scratchy dry ice inflicted on audiences in rock concerts. As soon as Marisol and I walked three steps, this mist enveloped us and the ordinary world disappeared, as if it never existed.

Day begins at 4.30am. On the ground, amid a russet and green carpet, leafcutter ants start their daily labours of nibbling, snipping, lifting, erecting scaffolds and dismantling them. The fer-de-lance viper, *Bothrops asper*, awakens from its cool-induced torpor and starts to unwind its car-tyre-thick body. Like a thermal imaging camera flicked on, the diamond shaped heat sensor between its eyes blinks to life.

In the canopy, howler monkeys call. Howler monkeys are black, demure-looking primates, until they open their mouths. If only they howled. Their sonic boom reverberates through the static pre-dawn air, carrying for kilometres through the forest, a primal and unnerving sound, like being summoned by angry ghosts.

The cloudforest's lack of depth of field, of much visual field at all, results in vertigo. At ground level a muffling carpet of leaf-litter is thick with the funk of humus. Two or three feet off the ground, a retinal knit of cobalt lianas tentacles everywhere, sieving out light. Bromeliads perch, stout and dripping, like crowned jewels tucked into each elbow of each tree.

The visual trick of the cloudforest is to appear tapestried and monotonous at once. It looks like everything; it looks like nothing. Only slowly does the human eye begin to unpick its embroidery: the bearded clumps of Spanish moss, prehistoric bryophytes, wild garlic, laurel, the *papayillo* (little papaya) tree. The neuronal network of lianas, spidering vertically and horizontally, draws not dissimilar shapes to those of blood vessels in anatomy posters. Once you can read the rainforest, it stops looking like a mesh of green and becomes a pattern. In fact it is not green, but has so many iterations of the colour it is equally hard to perceive them: the grey-jade of the moss, the emerald of the understory leaves, teal shadows, the rhododendron murk of the fan palm, its hand held open to the sun.

In the cloudforest most life takes place twenty metres above a human head. You spend your time with your neck craned to the sky, to a point where you end up with a crick. There, you see the spidering capillaries of strangler figs across the pale turquoise of the sky, smaller and more urgent than lianas. On the ground, you might glimpse the shadow of a peccary or hear the sawmill rumble of the puma, but the truth is animals are rarely seen on the ground in the cloudforest – it's too dark, too dense, and they move too fast.

Day after day, Marisol and I walked up and up, toward an ever-retreating sky, into a shroud of thickening mist. The reticence of the cloudforest closed in around us. As we walked deeper into this vacuum chamber, sounds slowly began to fill it, but in a manner that reversed the laws of physics, like solids filling liquids. A drawer suddenly shut, a bell instantly muted, a threading bronze sound not unlike a gamelan, bouncing back and forth, an uncanny Morse code which had been being tapped out by the forest unchanged, long before human ears existed.

'Bellbird,' Marisol said, referring to an endemic species. The three-wattled bellbird is retiring and rarely seen. Its call is the other-worldly sound we had been hearing since we entered the forest. 'Let's just listen,' she said. 'We might even see one.' Sure enough, a flash of iridescent chestnut appeared and then vanished, like a localised bolt of lightning. 'The male,' she whispered. 'He is the one we have been listening to.'

She was excited, suddenly. Her change of state, from strict, even dour, instructress to acolyte was startling.

'Did you see its wattles?'

'His what?'

'The male has three –' here she put her fingers to the bridge of her nose – 'they are like pieces of liquorice.'

I hadn't, but I knew that almost all the birds that live in the rainforest and cloudforest were fantastical on some level – the toucan with its banana-long beak, the resplendent quetzal, which looked like a flying Christmas tree. It is as if the rainforest has decreed that only outlandish beauty is allowed to flourish.

'See that?' Marisol pointed to one of the identical-looking trees just off the path. 'That's oak. That one is laurel. As you can see, it's covered with bromeliads, mosses, heliconias, orchids, lianas, strangler figs. That tree is a young ceiba, followed by papaya, rubber, and a cathedral tree. There's a wild teak, and that one a cacao. I'll give you a tree identification book when we get back to town. You can borrow it. Tomorrow you'll identify all these for me.'

Even though I did not know her well, it was clear that it wasn't an option to say, *ah, Marisol, you've been looking at these trees for two decades and me for two minutes.* I glanced at her. Her demeanour had softened. She looked at me – encouragingly, I think. Somehow she had decided I was not another entitled idiot from a wealthier country. 'Okay,' I said, and we carried on.

That evening back in my hostel I lay on my bed, exhausted. The hours had passed in a blur of information. I felt like I'd signed up for an immersive programme in a language of which I couldn't even speak one word. It had never occurred to me until that point that the land – any land – was a language of its own. For the first time I realised the problem was not that this language was unfamiliar, but that I could not hear it. There was too much ignorance between us. The human language is built for specifics and the land speaks in the severe glamour of merely being. I had an intuition: we might never speak to each other, or only when it was too late.

All Men Want to Know

In Central America rain turns out to be something else. *Rain* is the word for what happens in discreet, fey places like England, as in water falling from the sky.

A substance sluices down the window of my tiny eyrie room in Casa Ridgway, a Quaker guesthouse in San José, the Costa Rican capital. The window is narrow, rectangular, set so high I have to stand on my bed or a chair to see out of it. The so-called rain is not water but a torrent of liquid metal that turns the world pewter.

During the months of September to December, Costa Rica's Meseta Central, or central plateau, the rains run on a precise timetable. The sun rises at 5.25am and sets at 5.31pm. Until 11.00 the sun shines, fierce, inquisitive, a shard slicing through your head. At 11.15, clouds nudge their heads above the mountains' plumb line, like curious cows peering over a stone fence. The cows grow over the next few hours, exploding into towering puffy orchids of a quivering gunmetal hue. Around 1pm water beings to pour from the sky, insistent but not particularly heavy. Charcoal clouds cast an instant night. By 3.30 or 4pm anyone caught on the street had better know how to swim.

My clothes will not dry. My shoes grow mould and my previously invincible Norwegian Army leather backpack disintegrates. At night I pull a damp sleeping bag over me, chilled, even though the temperature never dips below nineteen degrees. By day I huddle in wet cafes in the university district, rivulets of water flowing from my umbrella across the floor.

For weeks now, I have been buoyant, frightened into a can-do state of capability by the monumental task I had signed up for. But since arriving in San José every day had felt like a Monday in January.

Usually I like rain. I am a contemplative person. At least in the temperate latitudes, rain has the same rhythm as thinking. But in the tropics its insistence blots out thought. I feel chastened by it, confused about my purpose.

Plus it is impossible to travel anywhere, or do any research. The Pan-American Highway is often impassable. Despatched to Costa Rica to write about its remarkable nature and thick biodiversity, I find myself trapped in the city. *The city is outclassed by its setting* – a phrase I read in a rival guidebook which turns out to be an understatement. San José is a concrete block dropped in a sea of diamanté and jacquard. Yet it turns out to be a good place to write. I might be drawn toward wildernesses by some inexorable tractor beam, but cities are always where I write best. In cities, as a woman especially, you are more invisible, less exposed. You do not need a reason for being there. In the countryside in Central America, any woman travelling on her own is suspect, as I was about to find out.

Paradoxically, only in the city would I be able to write about the nature of Costa Rica, although that is another misnomer: there was only nature; there, all human endeavour including towns and roads and bars and petrol stations looked fatally tawdry. The country's real heart was the black-gummed lips of volcanoes, their calderas brimming with jade lakes glinting in the rigid equatorial sun. Costa Rica is a thin country, draped by emerald forests, as if cloaked in thick, respiring jewels. The fact of a country that makes up only 0.03% of the planet's landmass but possesses five percent of the world's biodiversity explained its fervent commitment to generating life. So much life was packed into the tiny country in part because it straddles two oceans, but also because of its complex topography. Life is thick, the air laden with noises: crickets, insects, birds, bats, monkeys, strange cries in the dark, shufflings of green iguanas or giant red and blue hermit crabs. For someone from the temperate zones, it was monotonously, bewilderingly lush.

I felt safe in the city, but I had been sent to write about nature, so I would need to brave the rainy season soon. Each day my guidebook did not exist, Rough Guides was losing money and market share to their competitors. This pressure was issued in various ways from the publishers' office in distant Covent Garden, spewed from fax machines (no email then), on thermal paper that dissolved fast in the heat and humidity, like secret messages coded to self-destruct.

In Monteverde, the sun emerged. This was an event – their weather was as gloomy as Manchester's. The whole town came out into the muddy unpaved streets of the town, still dressed in Wellingtons and rain jackets, to squint at the sky.

We walked to the entrance of the reserve. Suddenly Marisol's face bunched up, a strange tension, as if she were trying to contain an internal explosion.

'What is it? I asked.

'It's always easier with women,' she said. 'The energy is different, calmer. Men always want to *know*.'

'What do they want to know?'

'Everything.' She'd performed a demonstration with her body, jiggling her arms, shaking her head, as if she'd been struck by a mild bolt of lightning. Then she did something I'd not yet heard her do: laugh.

'Where are your children?' I asked. 'Do they stay here with you?'

'Oh no, they're in San José with my mother. They don't like it here. It's too damp. There's nothing to do.'

'What about – where is… ?'

'Their father?' A sharp laugh. 'He got a scholarship. He lives in Arkansas. He met another woman there, another gringa.'

Another. With a single word, I was implicated. I was a gringa, and I reminded her of the woman her husband had chosen over her.

On the lip of the forest, she had stopped abruptly.

'Do you hear it?'

I strained. No bellbirds, no murmur of insects, no swish of leaves. She was frowning. Was this one of her tests, another I would fail radically, my ears tuned to the wrong kind of sound, schooled on the squabble of cities?

'I can't hear anything.'

She turned around. Without being beckoned, I followed her. We walked down the path, the wet rocks beneath our feet – more slippery, even slimy, than they had ever been before. I thought I could feel something gathering strength in our wake, a dark insistence that came from no one direction or place.

'What's the matter?' I asked.

We arrived at the small hut where the reserve workers checked entry permits and stamped people in and out. 'Strange,' the young man who we'd seen many days in a row now and whose name was Justício, Justice – a heavy name for a cheerful, round-faced youngster – said, 'no-one has come to walk today. You're the only ones.'

'Maybe today is not a good day to walk,' Marisol said.

Justício gave her a look. 'Why, will there be a storm?'

Marisol walked on without answering. I shrugged at him, in apology, and followed her. It was a ten minute walk back to Santa Elena, the village on the edge of the reserve.

When we parted at the entrance to my hostel, she finally spoke. 'It's important to listen, when the forest speaks to you that way.'

'What way?'

'It didn't want us there today.'

'How do you know?'

'Because it told me. Didn't you feel it?'

'I did,' I said, thinking of the dark feeling that followed us down the path. 'But I didn't know what it meant.'

I'd felt an unease, possibly before Marisol had heard the forest say something she both could and could not translate. It was like a small seed at the base of my spine. The sense of something retracting, of taking itself away from us.

Later, I sat in my bunk-bed hostel room and stared at its faux-tartan curtains (perhaps its owner was a displaced Scotsman). *All men want to know.* In her statement, so abstract and casual, lurked the brute fact of the will of man, its insatiability. It would not stop until it had conquered women and destroyed nature, because they are the same thing.

I knew somehow I would not see Marisol again, and this is what happened – we planned to meet the next day so that I could put a list of questions to her, but something came up. There were no mobile phones, then. She left me a message at reception. *I'm sorry, I have to return to the city.*

On my last night in the cloudforest I had a dream. Marisol and I entered the forest together, dissolving into its gunmetal shadows. In the dream the forest did not warn us. We became lost, then separated from each other, then we emerged from it without warning into the ragged suburbs of San José, which absorbed us without a trace.

Currents

I unearth the book, now covered in a pelt of dust, from my bookshelf. *No-one ever reads guidebooks twice, or for the prose*, is the guidebook-writer's joke.

I thumb through it. My own words speak to me in a different voice, not mine, but not someone else's, either.

I find what I am looking for: my account of Cahuita, a small village on the Caribbean coast of Costa Rica that was 'discovered' by tourism not long before I began my research for the first edition of the Rough Guide. By the time I got there, the village was a famed budget tourist hangout. All the gringos were tanned and lanky and carried surf boards. Everyone looked inordinately smug, padding down the sandy avenues of the town barefoot, flanked by loyal temporary dogs.

This paragraph is from my official portrait of the place:

Cahuita has become a byword for relaxed, inexpensive Caribbean holidays, with a laid-back atmosphere and great Afro-Caribbean food, not to mention top surfing beaches south along the coast. The sheltered bay was originally filled with *cawi* trees, known in Spanish as *sangrilla* ('bloody') on account of the tree's tick red sap - Cahuita's name comes from the Miskito words cawi and ta, which means 'point'.

And here is an extract from my notes during the five days I spent in Cahuita on my first trip:

Drunks at 8.30 on a Sunday morning; a German woman up Black Beach, sitting with a beer writing a letter to relatives at home no doubt telling them of sunny, hot, beautiful Caribbean breezes in reality pissing rain, cold enough so I wear my giant raincoat, wind wheezing like blunted knives. Another blond woman having to be led from Bar Vaz at 5 in the afternoon, supported by another blond woman, while a forlorn little blond girl around five years old splashes in puddles, straggling after her mother and the drunken woman. I keep seeing her little red rubber boots.

'Something terrible gets people when they come here,' a hotelier, a Quebecois man, tells me. 'They end up drunks, drug addicts, crazy people.' He means the Europeans who come here on the promise of a laid-back life. I can't help but agree with him, there is a feeling of – what? False values, as if everyone is putting on a show but not for strangers, or others, but for themselves.

Then there's the violence. Talking to hotel owners, they don't want to tell me at first but when I ask about the machete attacks – when they rob you here they hold a machete over your throat – they admit to them, and to the European women hooking up neo-colonially with black men, a phenomenon known locally as Rent-a-Rasta...

The town is picturesque, with its collapsing clapboard houses, but the grey-sand beach is littered with driftwood. Sharks are said to patrol its shores. You can't even walk along the shore – a grove of tightly packed coconut palms make the coast inaccessible.

I don't like weed or beer, but I can feel even so how, if I stayed one day too long, I too might have to be led through puddles back to my beachside cabina where there is absolutely nothing to do. Except drink, of course. You can always drink.

These two accounts of Cahuita, the official and the actual, align to an extent. In my full entry I mention the crime, the lassitude, although not the local sexual economy. I couldn't find a way to express it to readers that did not sound puritan, prurient, racist, or all three. But the truth was I had never seen Europeans drunk in rainy tropical beach towns at 8.30am on a Saturday morning, their neglected children amusing themselves by splashing in puddles.

Yes, the landscape was hauntingly beautiful, with its mercury sunsets and palms that leaned so bashfully over the ocean, but its human world had a rancid atmosphere. Was it my job to point this out? Rank exploitation – tax dodging, child sexual exploitation, prostitution, drug trafficking – as well as ordinary bad faith, was visible everywhere, once you knew how to look. If I wrote about it, who would it serve? Disturbing currents ripple underneath all places, particularly those located in countries with vast social inequalities. Sometimes those currents are inherent within the land: just as there are bad people, bad places exist, too. Even if humans have lost the inner ear we once had for such frequencies, we do sometimes pick it up. In me, it feels like a wave of sound, an echo disturbing the frictionless liquid of the matter that moves between and binds us. An ill wind, if you will. Like the episode with Marisol in the cloudforest, when the land sent us a warning, but only she could hear it.

The guidebook writer has to walk a fine line. On the one hand, you are selling the place for future consumption. On the other, you are trying to write as accurate a portrait as possible, so that those future consumers do not feel they have been sold a lie. The problem is revealed in the verb: whatever you do, you are a salesperson.

Being a guidebook writer was never going to be compatible with full disclosure. For a writer of any integrity, to be forced to turn away from the truth makes the whole project of writing null and void. I spent so many evenings on the road in Costa Rica, parrying pity from countless well-meaning waitresses and barmen, (*pobrecita, solita, dónde está su familia* (poor little thing travelling alone, where is her family?), trying to measure the distance between what I saw and experienced, and what I was allowed to write, for a commercial aim.

I hoarded the truth I encountered in many places for my fiction, where truth can float free, untethered to an extent from economics. In those first two years I spent researching, writing and publishing *The Rough Guide to Costa Rica* I somehow managed to write the bulk of a collection of short stories which drew on my itinerant life working in Brazil and Central America, published as *Nights in a Foreign Country*.

I did spend so many nights in strange places, waking alone in beds I would never sleep in again, packing my rucksack and slinging it in a bus or plane or rented car, contenting myself at night with a dinner of toast and cream cheese, picking spiders off my pillow and plucking crabs out of my backpack. Nights spent on the edge of belonging, intent on grasping an apprehension I could not quite define but which would lead me toward knowledge. Not self-knowledge, but a more geological truth: the dream of the earth remade and freed of a future we can no longer cure.

One October night I was on my own in Casa Ridgway. The rain sluiced down in cyclonic quantities, as usual. I sat at the table watching it soak the courtyard wall, thinking longingly about autumn, the season of industry I was missing out on at home, a time of new starts, of renewed energy after summer, a season of intents. I was about to go to some remote area – Greytown, in Nicaragua, and the mouth of the San Juan river, and was gearing up for the effort it would take, driving on flooded jungle roads and flying in tiny Cessnas. I knew it would be a lonely, tough trip.

I considered how Costa Rica had fulfilled its ambition to show the world how conservation could be done. It set aside twenty-five percent of its national territory and built a network of national parks and nature reserves. Its efforts to preserve its biodiversity, one of the most extensive in the world, came to fruition.

But what of the cloudforest? On climate change, Marisol had been ahead of the game. The rate at which plants and animals were becoming extinct was now a thousand times higher than before humans inhabited the earth, she told me on one of our long, muddy walks. While tropical rainforests cover around six percent of the earth's surface, numerically, their land and water-based species make up half of life on the planet. It follows that the loss of a hundred species a day will fall most heavily in the tropics. Globally, continental South America had the highest extinction risk due to climate change of any region on the planet (twenty-three percent, followed by Australia and New Zealand at fourteen percent). This was due to the diversity of the species found on the continent and the narrow range of habitat of many. As the cloudforest warmed, there was nowhere for creatures to go, apart from higher. Those who could not survive in the cooler air at the top of the canopy, or whose temperature range was exceeded by the warming atmosphere, would die.

The cultural critic Lauren Berlant has written that 'the present [is] an effect of historical forces that cannot be known fully by the presently living.' I remember everything I have written about Costa Rica only because I wrote it down in notebooks pock-marked by rain, the ink dissolved into puddles, so that reading them now is like deciphering hieroglyphs. I keep diaries not only to bear witness to that most slippery dimension, the present, but because I know how much of our lives we forget. But how much can we forget or bear witness to something as vast and dispersed as the Anthropocene – a cold, intimidating word for everything I have seen happen to nature in my very brief lifetime: the diminuendo of it, the subtraction and deletion and distortion. There are no words, but there are memories.

Time has condensed those years I spent in Central America into a single image: a young woman wearing Coke-bottle glasses, legs scarred by mega-mosquito bites, hands gripping the steering wheel as she weaves between potholes. The windshield wipers barely make a dent in the rain, which bounces off the Panamericana with such insistence that the drops ping from the tarmac back up to the windshield. On the CD player she has Marisa Monte on repeat, singing *não vai embora, não me deixe nunca, nunca mais*. The woman at the steering wheel expects that in the future some stability and emotional coherence awaits. She is aware of her obsession with knowing the world, the physical, geographical reality of it, but never suspects that she is a witness to and agent of its destruction. As the dark corridors of the palm oil plantations flash by her, she cannot know that she will undertake versions of this same mission, over and over, for the rest of her life.

PART IX

BOREAL

July 30*th*, 2021. El Port de la Selva, Catalunya, Spain. 31°C

By late July summer has thickened on the Mediterranean. It hasn't rained in two months and the land is parched.

Yesterday I walked the coastal path that hems this northern quadrant of the Costa Brava, called the Camí de Ronda. I spotted a small green lizard with delicate stripes down his body motionless on its edge. I bent down, thinking that it was dead. If it was alive, it was certain to be crushed soon by a less attentive passer-by. The lizard didn't move. I took out my water bottle and poured some onto the ground. The lizard lunged for it. I watched as his tiny tongue lapped it up before it soaked into the dry soil. In the end I emptied my bottle. The lizard drank it all, then disappeared into the rocks.

I headed back to the small flat I had rented, to wait out the heat of the day. As I walked my thoughts drifted back to February of this year when a colleague at UEA invited the writer Amitav Ghosh to speak on writing about climate change. It was in the depths of the winter lockdown, another of the endless Zoom events I was professionally bound to attend. But this one would be different.

On the screen Ghosh was priestly, dressed in a black turtleneck with a shock of grey hair. He spoke from a Verdigris-coloured room somewhere on the eastern seaboard, where he is an eminent professor of literature as well as a novelist. He was to discuss his book *The Great Derangement* (2016), which questions the contemporary literature's reluctance to engage seriously with climate change.

Ghosh told us how, in the great English social realist novels of the 1800s such as the work of Thomas Hardy, Charles Dickens and George Eliot, the machine of the industrial revolution was present in the very fabric of the narrative and characterisation. Novels by English writers published from 1825 until the end of the century drew dramatic vigour and conflictual energy from the rapid societal changes wrought by huge social forces, he said. Why was the same not happening with climate change?

He had an answer. 'In part, it is because western narratives are about individuation and ambition.' Ghosh argued that climate change is in fact the background of everything that happens. We just don't see it. That will take time. Only future generations will benefit from the rear-view mirror in which climate change will be infinitely refracted. 'People now – they see that they won't have the lives their parents had,' Ghosh said. 'They are increasingly angry, afraid, they won't have the stability, the sense that they can move upward… that life will get better. They don't even have any confidence the earth will be habitable.'

Someone in the audience asked him why governments, institutions and individuals are not doing more to fight climate change. Do they not have enough information? Do they not realise what is at stake?

Ghosh replied with a weary, almost spectral declension in his voice, 'I think they *know*. They know what will happen. There is no longer any real denial. But the vested interests are too great. Our generation – we have lived off the fat of the land,' he said. 'Now, people are crushed. They try to blame everyone. They place the blame everywhere. Immigrants. Sexual minorities. But the blame is with the power structures that precede and supersede capitalism.'

They know, they know, they know – for months to come the phrase will ring through my head. Ghosh had exposed the lie of how we were still behaving as if the failure to address climate change was because we do have enough, or the right, information, that if only we understood the implications and dimensions of the damage, we would stop everything we were doing, and face the catastrophe we have created.

I've been lucky, more so than I knew. I am alive in this brief juncture between mutually assured nuclear annihilation and the slow apocalypse of climate change. I've been so fortunate, to live a heliotropic life, leaving Heathrow in the endless night of December and emerging squinting into the light – in Cape Town, San José, Rio de Janeiro, Johannesburg, Mexico City, Nairobi; years of Christmases of potted casuarinas in shopping malls, baubles glinting in the sun, the hum and wash of the troposphere still in my ears. I knew the damage I was doing, just by boarding a tin can powered by Aviation Turbine Fuel. But my desire to know the world before I left it was too great; it easily overpowered my growing awareness of the carbon cost of air travel, the guilt and the shame. This is another tragic fact in the Aristotelian epic in which we are all cast as pages and servants: that to know the world, to really know it in three dimensions, means burning a trail through the troposphere that singes the planet. Even when I became aware of the effects of what I was doing, I did not stop.

Two weeks after meeting the lizard on the path in Port de la Selva, I am woken at 5am by shouts and an urgent banging on the door. As I rise from sleep to the surface world, I make out a single word: *incendio*. There is an ashy smell. I part the curtains. The mountain is lined by carmine. *Hay que salir*, the neighbours shout in the street to the sound of cars revving. We need to get out of here.

Quickly I dress and gather my passport, my laptop, on which I've been working on a draft of a book I call *Anthropocene Diary*. I can't remember if I've backed up the latest version of the file. My brain is sluggish with alarm. The ashy smell is now joined by one of charred pine.

I emerge into the dark street, unsure of where to go. *Away from the fire, obviously.* The boss-me voice wakes up to order the sensitive-uncertain-me around, as it often does in crises. I start out walking for the town, head-torch clamped to my forehead, as if I'm in the bush. It smells like Africa, come to think of it: the campfires we make in the early morning to boil water for coffee, the tinderbox fume of dry veld.

Behind me the red line has advanced toward the houses in the neighbourhood where I am staying. Some millionaire's house, a sprawling construction of two wings, like an albatross, is only thirty metres away or so from the lip of the fire. The sky above it turns purple.

Port de la Selva is the end of the road, terminating in the wild Cap de Creus peninsula, a national park. It's cul-de-sac-ness protects it from mass tourism. But now it's a trap. We can always take to the water, I tell myself. I will be able to hop on someone's boat and sail down the Costa Brava. I am not afraid, or not yet – that is the trick with fear, to allow yourself to feel its dark impress, but not be overtaken.

I join some of my neighbours on the pedestrianised esplanade along the sea, clutching my backpack on my lap. We sit on the stone wall that divides it from the beach. There is no moon, only the light of the fire, veining down the black quilt of the Pyrenees. The combination of the fire, the mountains, sky and sea creates a shadowy violet zone, like the hungry penumbra of a distant planet. We all sit there in silence, until dawn.

Around eleven that morning a police car drives by with a megaphone. They announce, in Catalan, that the fire has been brought under control and the road to the nearest town with a railway station, Llànça, should be open by the early evening.

The mountain behind my rented flat is scoured by black chevrons of charred grass. I sit on the terrace outside, raking it with my binoculars before turning them on the yachts and speedboats in the bay, the Los Angeles-style mansions built into the granite hills, the cars winding back and forth along the seafront, all the human world continuing as if it were not just nearly incinerated. Then I notice the silence. Normally the air is thick with the churn of crickets and birds. The air – dry, with a note of recent heat that feels fugitive, somehow, as if it is trying to dissipate itself – rushes past my ears.

I am so far away from where I started life, geographically, culturally, intellectually. I speak four languages I did not know existed when

I was growing up. More money graces my bank account in a month than the people who raised me saw in three years. I am not the same person who played for hours on end alone in the snowy skirts of a boreal forest in eastern Canada. But more startling is my sense that the planet on which I came to consciousness in those cold forever winters of the 1970s is not the same planet I am on, here in the blue embrace of the Mediterranean. That planet is gone. Where do we live now?

Boularderie, 46.2487°N, 60.8518°W

The Quickening

Snow covers their land like the pelt of a vast animal. The girl does not need a thermometer; her nose tells her how cold it is. Colder than minus ten and the air inside her nostrils condenses and begins to freeze. The speed at which this happens – within a single breath – tells her it is twice that temperature.

The dog's heartbeat thuds into her parka. They are wrapped around each other, huddling underneath a wooden veranda where the snow cannot reach. The snowstorm they have come out into, at night, is no match for him. She hears the sound of things breaking from time to time, above the wind.

Rustlings come from nearby. The under-veranda is home to racoons who semi-hibernate through the winter. The dog whimpers, which is out of character for him. He has also been misguided, had his coordinates mixed up, funnelled to the wrong planet – Planet Winter, Planet Rage. Eventually someone on her home planet will notice she is missing, but that could be years from now. Or it may never happen.

Does she stay under the step and risk freezing, or risk going back inside to the dark tornado of his anger? In an ordinary, literal sense her world is upended, because he is throwing furniture around – spinning wheels, small tables. The heavy stuff is too much for him, no matter

251

how crazy. The following day he will have smashed hands and a pulled *biceps brachii*.

What can create such anger? What can contain it? His rage seems like an ordinary topographical event, like the bluegrass mountain on the other side of the salt sea system, their clapboard farmhouse snug in its eye. Or the tapestry of spruce and pine that surrounds them, thick as any stockade.

She does not know how much time passes. When the dog starts to shiver, she takes them both inside.

The white ash that falls from the sky turns out to be liquid. It floats, serene, above their heads. But when it falls to the ground it congeals and seals them in its pleats.

So much falls that sometimes they have to crawl out of the second storey windows. He makes a slide for them. They slip out of the bedroom windows and start digging out chickens, who fluff their feathers in gratitude, and grumpy pigs, who huff and snort. The dog, who has spent the night shielding himself from the blizzard in the barn, howls from the inside. They dig him out too.

The ice world has an unnerving stasis. She has been born into a realm far more fearsome than the dreams that racket through her nights like trains. From somewhere books appear, about white queens and dark-haired virgins who are put to sleep in glass coffins and who miraculously awake after years, rejuvenated by cold. The books tell her that what is popularly called life is actually a zone wherein people speed heedlessly toward entrapments.

She is surrounded by stealthy presences. She glimpses them from time to time in the woods: tufts of ears, dark moving obelisks, two coal eyes, a filigree crown of bone and fur. She has no doubt that these creatures are the real captains of this realm and that her presence is merely tolerated. If they wanted, they could kill her. But oddly they never do.

Instead they regard her with onyx eyes that burn with a subdued, abstract desire. They seem to want to communicate something to her. She waits for them to open their mouths and for the air between them to arrange itself in sonar arabesques, for sudden bursts of revelation. When this does not happen, she searches their eyes for clues to the current which runs between them, muted yet apparent, like a language spoken just under the threshold of hearing. Their eyes are not unknowing, exactly, they perceive her, they perceive her perceiving them. They know they have been seen.

For a long time she speaks to the creatures she encounters in this silent language, sparks emitted across a void and which bounce away from their skins before they can scald. She is thrilled to be apprehended. She is a child, and so still an animal. The animals know this. Galaxial bursts of red and amber are emitted from the deer's corneas, from the rust iris within the eyes of the bear. No human eyes are capable of that expression, but she will search for it nonetheless, for the rest of her life.

Winter is a rival, an obstacle, a foe. The winter can kill them with one snap of its fingers. It is a drastic God, an ivory grove which they enter with resolve and dread. Snow is falling through her hours, she is in a flaky diorama, she is fascinated to see it evaporate on the dog's tongue, she builds tunnels and snow forts and her grandfather teaches her to build an igloo using her lunchbox as a snow-mould. Their world is shaped by seracs and cornices, by long waving sentences of sastrugi written by the wind into bleak, stern patterns. The same substance will sculpt her inner world, too. She does not know that the white ash is not a forever world. She does not know, as David Wallace-Wells, an American writer, will put it fifty years in the future, that 'The climate system that raised us, and raised everything we now know as human culture and civilization [would become], like a parent, dead.'

As for ice, they are always falling through it. Men, women, dogs, horses, disappear through its lace meniscus as into wormholes to an adjacent dimension. One of the first things she learns as a child is to fish a person or an animal out of its grip.

They are out in the woods, chopping firewood, when they hear a sharp spectral howl.

'Come on,' he tugs at her parka. 'Put on your snowshoes.' He looks like a lumberjack in children's books: red and black checked shirt, fur hat with flaps to cover his ears, mukluks, a still-warm pipe jutting from his mouth.

'Where are we going?'

'The lake. There's an animal in trouble.'

There is a slight swishing sound as they move. The air stings her cheek like knives, but it is pleasant, this slicing sensation. As they near the lake the howl becomes sharper.

The animal is a strange squiggle on the ice. Half of it is missing.

He takes off his snowshoes and whips out a coil of rope from his backpack. She finds herself being pushed backward and this too is weirdly pleasant, to not know where she is going, what is behind her.

'What are you doing?'

'I'm tying you to the tree. Here keep still.'

'How will I get free?'

'It's a figure of eight knot. You just pull the long cord.'

Rope goes around her hips, tighter and tighter. The resin-mint smell of pine. He then spools the rope around his hips, once, twice. She is tied to the tree, and he is tied to her. Why not just tie himself to the tree, she wonders? Perhaps he is protecting her. He knows that if anything goes wrong, she will rescue him and the wolf, in that order. She will not stop trying until she is also dead.

She closes her eyes and clasps her fingers together in her mittens to stop them from freezing. She watches him disappear across the ice, sliding toward the wolf on his stomach.

She can't quite see what happens next. Somehow the wolf does not savage him but allows itself to be clasped in his arms. Still on his stomach, so to better distribute his weight over the ice, he pulls the wolf's hind legs out of the water. With the grey-black animal still in his arms he calls to her.

'Pull!'

'What?'

'Pull me back!' His voice is rough, like a saw.

She tugs on the rope with all her strength, winding it around her mittens. She can't pull a man and a wolf on her own, but something of the traction she creates eases him to stand up, to gain his footing and then he is walking with a full-grown animal thrown over his shoulders.

Another instruction, as the man-wolf figure walks toward her. 'Start a fire.'

They have practised this many times, her lighting a fire with him watching. In case he is injured on their many expeditions in the woods, he has taught her to cook, to navigate, to raise the alarm with the flares he carries.

The wolf is warmed and its hind legs saved, it stays with them for three hours as they sit round the fire. Her grandfather sits on the emergency blanket he carries rolled up under his backpack and feeds it pemmican. The wolf's eyes are the colour of haybales. It closes them heavily, exhausted from its ordeal. When it finally leaves them it does so in chapters: trotting away a few metres, looking over its shoulder. Then a few more, repeating the gesture. It seems to be reluctant to leave.

Once, there is a fawn fallen half-through the ice, just as the wolf has done. With its matchstick legs it scrabbles and scrabbles but cannot pull itself out.

This time he winds a rope around her, removes her snowshoes, and shoves her out onto the ice. The fawn's mother frets on the perimeter of the lake. The girl shimmies out on her stomach, worried she will tear her favourite red snowsuit with the Inuit stitch pattern. The fawn is only half through the ice, its forelegs dangle in the water. Its little body is surprisingly sinewy, like rope covered in the thinnest layer of skin. It shivers uncontrollably. She clasps it to her, both of them trembling, and her grandfather hauls them back across the half-frozen lake.

He takes the fawn from her, brushes the snow and ice off its legs, and tucks it inside a rough woollen blanket. He rubs and rubs the fawn, ignoring the alarmed plea in its eyes, and they both blow hot air on its nose. The fawn closes its eyes against this onslaught. He dries its legs with an old towel usually put to service cleaning the rifles. It smells of gun oil and copper – a smell that will arrest her almost four decades in the future as she cleans her own .375 Holland & Holland, in the far north of South Africa.

With the fawn bundled in the blanket he snowshoes toward its mother. He deposits the fawn in front of her like a package. They retreat to a safe distance and peer behind a tree to see if the mother will accept it, now that it has been cloaked in human smells. If not, they will be taking the fawn home, despite the protests of her grandmother, and it will grow up in their kitchen like an outsized puppy. She half-hopes this will happen, but the doe noses the fawn to its feet and begins to suckle it. They stay for an hour, until they are nearly frozen to the bone, watching, to make sure.

Ice is always treacherous, she learns, especially in early spring or late autumn, swing seasons when the ice looks thick but is actually dissolving from underneath. None of them – the ageing man, the child, the fawn – can hear it yet, but far underneath the lake the pulse of the planet is quickening.

Latitudes

The province of Nova Scotia juts out into the Atlantic Ocean like a fish turned on its side. From above, the land appears to be more water than terrain; there are 6,674 lakes in a landmass roughly the size of Bosnia and Herzegovina or Costa Rica. Five bodies of water surround the province: the Gulf of St Lawrence to the northwest, the Northumberland Strait due west, the Bay of Fundy to the southwest, the Gulf of Maine to the south, the Atlantic Ocean to the east.

Cape Breton Island began life as the Amazon rainforest. The island's seams of coal, which were once its sustaining and singular lucre, were brewed of compacted *Lepidodendron* and *Sigillaria* trees which last saw the light of day in the Carboniferous period from 359.2 to 299 million years ago. The island migrated north on the Gondwanaland continental breakup. Along the ride it swapped scarlet macaws for mastodons, whose disappearance not coincidentally coincided with the arrival and establishment of Clovis Hunters, migrants from Asia who traversed the Bering Strait in the final days of the last ice age around fifteen thousand years ago and spread rapidly throughout the new landmass.

Cape Breton is on the same latitude as Lake Como in Italy, but they may as well be in different hemispheres. Cape Breton has no access to the Gulf Stream, a thousand kilometres to the east. The dominant climatic regime is Maritime Boreal, characterised by warm summers, cold winters and moderate rainfall. Long, cold winters and short, cool and mild summers are optimal spruce growing conditions, and Cape Breton has plenty. In places the spruce grows so tightly packed even a small child cannot penetrate it.

I have always loved the word *boreal*, meaning 'of the northern regions'. The word is derived from Latin, but as is the case with many words from a seemingly Latin origin, there is a back door in their history which opens onto a mysterious past. The root is thought by most lexicographers to be the Greek word *Boreas*, which means 'God of the North Wind'. But there is another possibility; some dictionaries say that it began life as ancient Balto-Slavic for forest, or mountain. The forest where the God of the North Wind lives – that is just how I would describe the reticent garrisons of spruce, the cold, cobalt lakes.

Boreal forests are hard places to make a living, sodden as they are with bog-woods, or *tamarack* in the Mi'kmaw language, where vast quadrilles of mosquitos, deer flies and horseflies breed. Interminable winters are followed by short, devouring summers with no decompression zone between them. Like a bear, the land slumbered for much of the year, then woke with a violent, confused start. More than a season, winter was a way of being, a complete reality which demanded fealty. If you tried to reject it, it would find a way of claiming you back.

I wonder, could such a volatile climate imprint its nature on those who come to consciousness within it? The people I come from felt intensely, were prone to abrupt emotional swings. They brooked no compromise and were almost supernaturally tough. Their emotional template was the same as our woodcut world: black, white, white, black, the lapidary winter skies lit by a piercing, phosphorus light.

I had no father and no mother and perhaps I adopted Cape Breton as a surrogate parent. Cape Breton is an orphan sort of place, geologically neither wholly of the Canadian landmass, but also it has little of the Norwegian rock of Newfoundland, the larger island 110 kilometres to the north, with this Fennoscandian shield. The sun roamed the sky, untethered. It was a place of stone days, of black rebellious clouds patrolled by bald eagles twice my size, with their eerie twilight eyes. A great glacial stasis remained within its memory. When it did speak, its voice was low, booming, bass. It must have been in those long hours when I started to hear it, in the same moment my own voice began to ricochet around my head.

What is a landscape after all but an ideology, a thought-system? This must sound grandiose, I realise. Surely the land is just that: land. To force it to play any other role is just hubris. Humans habitually twist things that have no interest in us to bend to our consciousness, or worse, our vanity. Yet I can't shake the thought that the land is family. Even better than family, it is our accompanier, a friend, a fellow traveller met along the way, even if it remains static while we are forced to move on.

Degrees of temperature, degrees of distance from the Equator, the Poles: we have lived our lives within a grid of expectation, latitude and longitude and the thermometer we are all trapped in and which moves only in one direction. The planet is sliced and diced, in order that we may find our way around it, via these degrees of separation. We rattle around the earth on their carousel, twenty, thirty, fifty, zero degrees, thinking that we know this earth, that latitudes give us – there is no other word – latitude to chart and so govern and control and subdue and rob.

My suspicion is that we don't quite believe it, that we are here, that this planet is our only life support system, that there may be no other like it in all of infinity. Somehow within our minds a delusion operates: once we are through with this place, once we have filled our Coutts accounts and bought yachts and sovereign funds from Brunei we will simply find or make another.

Meanwhile, the coordinates where we come to consciousness imprint themselves upon us, as effectively as any astrological profile. We are date-stamped by them. My barcode is 46.2487° N, 60.8518° W. Even if I never see Cape Breton island again, it will accompany me until my death, and possibly after. I have internalised its cornflower skies, the artichoke-hued forests, a green so dark they appear black, the fast-deepening light of its sudden springs. The land is me and I am it.

As for the people who brought me up and taught me to survive in the boreal world, I see them, now, clearer than when they were alive. Why is it that we only come into focus once we have vanished?

They are poor people, married young. Their ancestors did not choose to come to this new world, they were deported by the English, who wanted the land out of which they eked a living for sheep grazing, so they too could fill their Coutts accounts to the brim. The new land nearly killed my ancestors; only by the intervention of the Indigenous peoples, who had been living on this inhospitable island for at least thirteen thousand years and knew its convulsions and its occasional bounty, did they survive.

In all the photographs it is either winter or summer; the swing seasons seem not to exist. My grandparents are dressed like twins, they wear matching lumberjack shirts of red and black check. They are taking a break from shovelling snow, which accumulates around them in tors and seracs of a stature I will not see again until I travel to Antarctica. My grandmother's hair is tucked under her felt hat. She is smoking a pipe. My grandfather looks demure by comparison. My grandmother is the man, in her corduroys and mukluks, the pipe jutting from her mouth. The snow makes rigid silhouettes out of both of them as does the zigzag of spruce trees standing sentinel around the house, dark blots on a blare of white. They stopped existing thirty years ago now, but I wonder if the land might remember them, the brief incision they made into its tissue, before it moves on, into the vertigo of time.

Epilogue

July 26ᵗʰ, 2021. London, UK. 24°C

I find my way back home to London in the second year of the pandemic, after three months away, first in Kenya, followed by a bout in Spain to dodge the UK government's racist and pointless 'red list' hotel quarantine.

At Heathrow posters of Beefeaters and exhortations to consider Britain 'Great' meet me. My wariness at living in Brexit Britain has been transmuted into perma-disgust, to the point I feel nauseous, all the time. But I still love London.

The city is wilder than you think. With three thousand parks alone – the most parks of any major European city – and a network of canals, reservoirs, rivers, wetlands and urban gardens, twenty percent of the urban space is given over to nature. Red kestrels patrol the marshes where I run each day in the Lea River Valley. Yesterday a heron and I did a hopalong pas de deux; the heron standing on the path on his stilts, me running toward him, him taking flight and landing further along the path, repeat, repeat. Adders, with their tar black bodies mottled with yellow, coil in the long grass on the marshes, as do swans, Canada geese, even a pair of Egyptian geese who forget to migrate each winter. Blackbirds, chiffchaffs, yellowhammers and rose-ringed parakeets start singing at 4.30am each morning at this time of year while an army of urban foxes plies the nocturnal byways of the city, sleeking through gardens and ruffling rubbish.

I find myself thinking about these last few years, casting back to the conversation at Malcolm's at Cley, in the unrelentingly hot summer of 2018 when it did not rain for fifty-five days. It was the moment zero of this book, although I can only see that now.

Like everyone, I am trying to adjust to the Anthropocene uncanny. It is too hot, too cold, too much rain, not enough rain, no snow, too much snow, too windy or weirdly dead calm. The world has come off its wheels, like that fairground ride I so loved as a child, the Tilt-A-Whirl, which swung you round and round until the gums started to detach from your teeth, like astronauts-in-training in a centrifuge. Winter in autumn, summer in spring, spring in winter, summer in autumn. All over the world people have the same feeling, an alloy of *dread-fear-guilt*, a new kid on the emotional block.

For a long time I thought all my travelling and throwing myself into totalising environments had been in search of the truth climate change denialists were not acknowledging. My desire to know the world was driven first by the impulse to escape the restrictions of poverty, then by environmental concern and interest in climate change science, with a slice of adventurism thrown in. But over the past twenty years a note of requiem has stolen into my work, until now it is the only note I can hear.

I see the problems, in facing up to the Anthropocene. Most obviously, writing about it is so much fiddling while Rome burns. But there are two failures beyond this. We can't seem to tell the story of our own self-destruction, even as we enact it in real time. This is not the tale we want to hear; it is, in the words of my friend, the Malaysian-Australian writer Beth Yahp, an unlovable story, as well as an unliveable one. As a species we are hooked on the same story, one of heroes, personal triumph, redemption, individuation, self-justification. It seems to be the only story we can write, even if it is killing us and permanently altering everything that sustains us.

But also, we don't seem to have any other language for the world-that-is-not-us apart from one of self-gratification and value. This language is an artefact of the last five hundred years, at least in English, which means England: invasion, tribal dynastic monarchy, imperialism, absorption of resources, slavery, mass deportations (the Clearances), industrialisation, mining, and now, latterly, a kind of accelerated, nihilistic land grab of the remaining resources we have, before the drawbridge is pulled up. From all these processes the words *resource* and *value* have been born, mutant twins. If we are to understand the non-human world we need a new conception, a new language, which recognises its autonomy.

Home at last in London, I try to rediscover a rhythm and purpose with my writing. I must finish this book, sooner rather than later, because as climate change swings faster on its deadly axis it is already out of date, even as I type this sentence.

I wonder again why I feel more connection and communion with places than with people. No doubt my lonely childhood, in a remote landscape with no other children or friends, is the diagnosis. But it is not the whole story. All my life I have held myself apart, fascinated by people, at least enough to become a writer, but in an anthropological way, as if I am doing a PhD on the customs and worldview of a creature so distinct from me we may as well be another species.

Over the last three decades, gradually at first and then with a sickening urgency, I have realised the import of what we are doing, merely by existing, how we are a death wish on the planet, and with this understanding my zeal to live in the world and comprehend it on its own terms, rather than the human, grew. And as this quest advanced, my relationships with people, always problematic, painful, perplexing, withered. Yes, I could write all this, offer it as an explicatory, consoling narrative. I just have. But somehow this is not the truth, or at least the whole story, either. That is because the story is in the land and not in me. I am only a transient interpreter of its silences.

As for *story*, for me, the word has come to have the same charge as vodka or crystal meth. In this book I've skated over or not written at all about many experiences I've had on these journeys, mostly bad: bad faith, bad relationships, bad decisions. All of this *Sturm und Drang* would have had the salty tang of the human. It would have made a good story. I was tempted to dramatise the weaselly letdowns, naked power plays, my own inability to defend myself, my intellectual arrogance, my inconsistency, my knack for being attracted to impossible people, my instinct to idealise place, the restlessness and tendency for fantasy which must surely be a symptom of a cowardly inability to commit to being myself, to commit to life, even. But the truth is – as a writer friend of mine who is struggling with a memoir has put it – impossible to write. 'Can you imagine what would happen if you actually wrote the truth?' she says. 'Nobody would believe it; nobody would speak to you ever again.'

Outside my window the estuary sky folds itself away as if constructed in panels, like shutters. It starts to rain, so hard it seems to stop time.

England has some of the oldest daily weather records in the world, some of them have been kept for six hundred years. London has flooded twice this month – on July 12th and again yesterday, on the 25th. On both occasions a month's worth of rain fell in a few hours. The city's Victorian-era sewer network was overwhelmed. People in basement and ground floor flats were invaded by a muck of water and sewage. Streets in my neighbourhood became sudden lakes. From my desk I watched cars hydroplane, their drivers taking the curve in the A10 outside my window too fast.

The feeling of unravelling is now everywhere. There is no place on earth where people are not affected. The average number of weather- (read climate-) related disasters has increased almost thirty-six percent since the 1990s. In today's *Guardian* George Monbiot writes that with present fossil fuel consumption we are on track for 3.9 degrees of warming. It seems clear that we will need to be pushed over the edge to finally address climate change. But by then it will be too late. Then we know what will happen: the rich will persist in their bunkers and bubbles. They may survive. But surviving is not living. Meanwhile we do not know how to grieve for the world, or to atone for what we have done to it. There are no rituals, healing ceremonies, other planets to migrate to, no Control Z key we can hit to reset to 1955 or 1860 or 1619.

I wonder, not for the first time, what would it feel like not to have the Anthropocene hanging over our heads. A future I am not afraid of. Species multiply and prosper. Bees charge thick through the air. Ice tongues loll over oceans, cooling and solidifying. The Himalayas accumulate. In the earth rhizomes invest in cadmium, burnishing trees into existence. There is a web and we are in it but we do not dominate. Each night tilts toward a dawn which will not be warmer than the dawn of the day before.

The impossibility of this vision sinks into me, like someone pressing too hard on a bruise.

Diaries are useful because they record the flavour and texture of that fleet animal called reality before it is converted into history.

For the record, this is what is happening now: the afternoon light has leaked from the room. Streaks of lavish sun are replaced with blocks of palest ermine, so watery it becomes translucent. The summer solstice is past; the light is leaving the British Isles, turning its back on us, or is it the world abandoning itself? Still, time rotates sturdily, keeping its valence in a vacuum.

I scroll back in time, through the file titled *Anthropocene Diary*, to find this entry from the previous year:

January 31ˢᵗ, 2020. London, UK. 11°C

Britain is unpeeled from the European Union, 3.5 years of hell have their apotheosis. Peak temperature on the Antarctic Peninsula: 18. A psychologist friend says, 'Much like the planet, people have a tipping point.'

Meet with my writer friend J., confess I no longer have the confidence to write, I am overwhelmed by events. 'Writers can bear witness,' she says. 'That is what we do. The world may be falling apart but we at least have a purpose.'

I go running on Walthamstow Marshes. Suddenly I am engulfed in a flock of rose-ringed parakeets. No-one seems to know how they came to colonise London, but they thrive in our newly mild winters. They arrow a metre or two above the ground, horizontal green rain from a magnetic dimension. I feel a small displacement, wisp of a breath as they bank, tilting like fighter jets, making room for my body. Their wings brush my arms, the thinnest gust of air. Then they are gone.

Acknowledgements

This book covers a lot of ground and time – indeed nearly a lifetime. My heartfelt thanks are due to the organisations who helped make the research and experiences that underpin it possible: the British Antarctic Survey, the Natural Environment Research Council, Arts Council England, Canada Council for the Arts, Shackleton Scholarship Fund, Galleri Svalbard, Helena Chavarria and Camino Travel in Costa Rica, the Environment Institute at University College London, A Rocha Kenya, the Sydney Social Sciences and Humanities Advanced Research Centre at the University of Sydney, the Humanities Research Centre at the Australian National University. Special thanks go to my home institution, the University of East Anglia, and my students there, who have taught me so much over the years.

My thanks are also due to my literary agent Veronique Baxter at David Higham Associates in London, and to Martin Goodman and his dedicated and professional team at Barbican Press, especially Mike Hart for his attentive editing.

I owe a debt to the work of the climate change scientists, journalists and writers (sometimes one in all at the same time) whose work has been so critical to advancing our understanding of our complex predicament, among them Amitav Ghosh, Mark Maslin, David Wallace-Wells, Bill McKibben, George Monbiot, the late Barry Lopez, Elizabeth Kolbert, as well as the many patient scientists I have met over the years on my official expeditions, who took time and care to educate me in their disciplines.

This book is dedicated to my friends, whose support and influence is the bedrock on which this story of spirit and the land rests. They have inspired and sustained me in so many ways, through their literary and artistic work, their comradeship and hospitality: Mark Cocker (for his reading and suggestions on an early version of this book), Julia Bell, Margie Orford (who read a draft of *Latitudes* and gave me her valued advice), Henry Sutton, Andrew McNaughton, Steven and Denise Boers, Nick Dennys, Christy-Ann Conlin, Rupert Thomson, Kate Norbury, Emily Perkins and Beth Yahp. The people who looked after me in those forever winters of long ago are the only reason I exist: John Charles, Christine and Janet McNeil. And finally to Diego Ferrari, for being my family.

Milton Keynes UK
Ingram Content Group UK Ltd.
UKHW022100121124
451112UK00011B/230

9 781909 954113